MANAGEMENT OF IRRIGATION AND DRAINAGE SYSTEMS –
A SERVICE APPROACH

UNESCO-IHE MONOGRAPH 3

Management of Irrigation and Drainage Systems – A Service Approach

HECTOR M. MALANO
*Department of Civil and Environmental Engineering,
University of Melbourne*

PAUL J.M. VAN HOFWEGEN
*International Institute for Infrastructural,
Hydraulic and Environmental Engineering, Delft*

CRC Press
Taylor & Francis Group
Boca Raton London New York

CRC Press is an imprint of the
Taylor & Francis Group, an **informa** business

CRC Press
Taylor & Francis Group
6000 Broken Sound Parkway NW, Suite 300
Boca Raton, FL 33487-2742

© 2006 by Taylor & Francis Group, LLC
CRC Press is an imprint of Taylor & Francis Group, an Informa business

Visit the Taylor & Francis Web site at
http://www.taylorandfrancis.com

and the CRC Press Web site at
http://www.crcpress.com

Table of contents

Preface

This monograph is written with the aim of providing an overview of the principles necessary to develop a service orientation in the management of irrigation and drainage systems. The material covered in the book is designed to address an area largely neglected in the irrigation and drainage management literature.

This neglect was the outcome of policies developed through the 1980's to transfer irrigation and drainage system management out of the direct government sphere, in the belief that such policies of privatising these activities would inevitably lead to improvements in the sustainability of irrigation and drainage systems and agricultural production.

The message contained in this monograph is that the principles of service oriented management need to be embraced by statutory authorities, public corporations and private organisations alike if these improvements are to be achieved. Simply "rearranging the deck chairs" is not likely to achieve significant improvement in irrigated agricultural productivity, and will not meet the broader objectives of integrated water resources management.

The philosophy embodied in this book is that irrigation and drainage systems must be managed as a service business responsive to the needs of the changing requirements of its customers. These are chiefly farmers in the case of irrigation, but may include urban and industrial areas in the case of drainage systems.

We believe that this service approach to the management of irrigation and drainage systems constitutes a key element of the strategy that is needed to improve the existing level of performance of many existing systems. Enhanced performance is a prerequisite if we are to face the enormous challenges to produce more food to meet the demand of a growing world population in an environment with an increasing competition for water.

This is a *why book* that emphasises *what should be done* and *who should do it*. It is not a *how to* book. It bridges the classical engineering and hydraulic principles related to irrigation and drainage engineering with the basic management principles related to the management of service organisations.

The book is written to serve the needs of managers and planners of irrigation and drainage systems as well as students of irrigation, drainage and water resources management. The material presented contains both a synthesis from relevant literature as well as conceptualisation of the several service principles applied to irrigation and drainage. Most of the literature on service management has been developed for commercial and industrial applications; and although the concepts are largely applicable to the provision of irrigation and drainage services, a large amount of adaptation was needed.

The content of the book builds largely on the teaching and research efforts of the authors who have delivered this material for several years to middle and senior irrigation managers and post-graduate students at the University of Melbourne, Australia and the International

Institute for Infrastructural, Hydraulic and Environmental Engineering, (IHE-Delft), The Netherlands.

Chapter 1 provides a perspective of the historical role that irrigation and drainage have played in global food and fibre production and the future challenges facing this sector. The chapter provides a rationale of why improving irrigation and drainage system management is imperative to realise the full potential of irrigated agriculture and increase productivity in a sustainable way.

Chapter 2 focuses on the external environment surrounding the management of irrigation and drainage organisations. It highlights the nature of water resource as a common pool resource and the underlying principles of Integrated Water Resources Management. It also provides a perspective on the different and sometimes conflicting interests between the main actors in irrigation and drainage — The Government, The Irrigation and Drainage Organisation and the Users.

Chapter 3 focuses on the internal management environment of the irrigation and drainage organisation. It explains the basic principles of management and strategic planning and its role in the management of irrigation and drainage services with a service orientation.

Chapter 4 elaborates on the principles and elements of irrigation and drainage services. It discusses the main factors that determine and conditions that must be considered in the formulation of level of service specifications. It also discusses the need for and different types of accountability levels and mechanisms needed to ensure a service orientation.

Chapter 5 discusses the key relations between the level of service and necessary hydraulic control and management inputs needed for the provision of a specific level of service. The various types of flow control concepts are discussed together with the associated operational requirements. This provides the basis for establishing the critical link between the level of service and cost of service.

Chapter 6 provides the basis for the sustainable management of the irrigation and drainage infrastructure assets. This approach is designed to provide a methodology that would enable the irrigation and drainage organisation to ascertain the actual cost involved of providing irrigation and drainage services. The framework and methodology for implementing an asset management program are also discussed in detail.

Chapter 7 discusses the assessment of performance of irrigation and drainage systems as the closing link in the management framework of the organisation. The internal and external performance domains to the organisation are outlined although the focus stays with the assessment of the organisation performance in relation to the delivery of services.

Chapter 8 summarises the potential and opportunities to introduce service oriented management as a common response to the many challenges the irrigation and drainage sector is facing. It emphasises that service oriented management cannot be looked into in isolation but should be seen the wider context of water resource use and management.

The format adopted for the presentation of the material includes tables, figures and boxes. Tables and figures are used to assist with the explanation of concepts and principles; whereas boxes use used primarily to provide examples relating to actual irrigation and drainage systems and to emphasise key principles.

We hope that what is presented here will play a small role in promoting the shift in paradigm in irrigation and drainage management towards service orientation to overcome the poor performance of many irrigation and drainage systems in past decades.

Acknowledgements

The generous assistance given in the preparation of this book is by many colleagues, associates and friends is hereby cordially acknowledged. We owe special thanks to Hugh

Turral for his contribution and useful comments. For generously sharing his research results and insightful comments on the draft we owe a special debt to Bart Snellen. We would also like to acknowledge with thanks the inspiration and foresightedness provided by David Constable who pioneered many of the concepts expounded here. Also we owe our thanks to Henri Tardieu for his valuable comments on the draft and Bart Schultz for both his valuable comments and facilitating the production of this monograph. Last, but not least, let us thank all those who helped us with the editorial and logistic support throughout this work, especially Fiorella Chiodo and Peter Stroo.

Comment and criticism from the readers will be sincerely appreciated.

Melbourne, Australia   Hector M. Malano

Delft, The Netherlands   Paul J.M. van Hofwegen

Abbreviations and Acronyms

CBD	Cost Based Depreciation
FAO	Food and Agriculture Organization of the United Nations
FMIS	Farmer Managed Irrigation System
ICWE	International Conference on Water and Environment, Dublin 1992
1DB	Inter-American Development Bank
IDTC	International Development Technology Centre
IFPRI	International Food Policy Research Institute
IHE	International Institute for Infrastructural, Hydraulic and Environmental Engineering, Delft ,The Netherlands
IWMI	International Water Management Institute
IWRM	Integrated Water Resources Management
O&M	Operation and Maintenance
ORMVAM	Office Regionale de Mise en Valeur Agricole de la Moulouya
SAR	Societe d'Amenagement Regional
SCADA	Supervisory Control and Acquisition of Data
UN	United Nations
UNCED	United Nations Conference on Environment and Development, Rio de Janeiro 1992
WUA	Water Users Association

Chapter 1

The Context of Irrigated Agriculture

The management of an irrigation system has for its purpose the delivery of water to agricultural lands at such times and in such quantities as will enable the irrigator to produce the largest and best crops. The success of the manager is largely measured by the success of the farmer.

F. H. Newell, 1916

Irrigation and drainage play an important role in global food and fibre production. However, growing water scarcity, inappropriate management of water and irrigation and drainage infrastructure, and declining soil fertility in many regions of the world are beginning to constrain future increases in output. The future increase in food demand has to be satisfied by making better use of present irrigated area as the potential for expansion is limited by the availability of and the increased competition for good quality land and water. Improving irrigation and drainage system management is therefore a necessity to achieve the full potential of agriculture and increase productivity in a sustainable way. Moreover, management reform is imminent because of increasing competition for public funds.

1.1 THE IMPORTANCE OF IRRIGATION AND DRAINAGE

Over 275 million hectares of land are currently irrigated in the world. There has been a steady increase in newly irrigated land of the order of 4 million hectares or 1.5 percent per year over the last decades. In addition, some 150 million hectares is provided with a drainage system only. The new schemes were responsible for a remarkable increase in the production of rice, wheat and other crops. Figure 1.1 illustrates this trend in irrigated area. For many countries this development resulted (temporarily) in self-sufficiency in food crops and some countries even developed an export capacity. However, this achievement is threatened by a growing population, a larger per capita consumption, a degradation of the available land and water resources and a heavier competition for use of these resources.

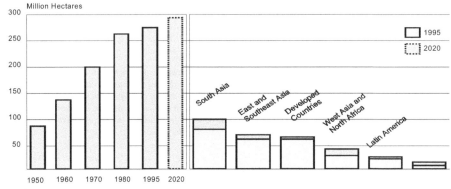

Figure 1.1 Evolution of irrigated area 1950 – 2020 (source: Plusquellec 1988, Rosegrant e.a. 1997)

1.1.1   Population Growth

The world population is now almost 6 billion and is expected to increase at a rate of some 80 million a year during the next quarter century, increasing world population by 35 percent to 7.7 billion in 2020 (United Nations, 1996). The low population growth projections indicate a stabilisation of the world population at around 8 billion over the next 40 years. Medium and high projections expect the world population to grow to 10 and 12 billion people respectively by the year 2050. More than 95 percent of the population increase is expected to occur in developing countries, whose share of global population is projected to increase from 79 percent in 1995 to 84 percent in 2020. Over this period, the absolute population increase will be highest in Asia, but the relative increase will be greatest in Sub-Saharan Africa, where population is expected to almost double by 2020 (Pinstrup-Andersen *et al*, 1997).

1.1.2   Per Capita Annual Renewable Water Resources

Unless properly managed, fresh water resources may well emerge as the key constraint to global food production. The world annual per capita, internal renewable water supply is about 7,000 cubic meters at present. This per capita supply varies greatly between and within countries and seasons. In some countries like in the Middle East and North Africa, the annual per capita supply is often less than 1,000 cubic meters. With a continually growing population, per capita water availability is declining steadily. The UN's 1994 medium population projection suggests that by the middle of the coming century, 4.4 billion of the nearly 10 billion people projected to inhabit the planet will live in 58 countries experiencing either water scarcity or water stress[1] (Population Action International, 1995).

---

[1]   Water scarcity is defined as less than 1,000 cubic meters of renewable water per capita per year. Water stress is defined as available renewable water between 1,000 and roughly 1,700 cubic meters per year (Population Action International, 1995).

The overall amount of water resources diverted for various uses is low. On this basis, it would appear that there are no major water resource constraints to further development of new irrigation areas. Global figures however tend to mask the marked regional differences in water resources availability. The reality is that there are many river basins in the world where water resources development has reached nearly full utilisation like the River Nile. Others such as the Mekong Basin in Vietnam, while being in an overall favourable water quantity situation, have limited capacity for agricultural expansion because of temporal variability of flows unless supply augmentation is pursued by additional regulation capacity.

1.1.3   Increasing Competition for Fresh Water

Water use by irrigated agriculture is very high compared with other uses. Worldwide, approximately 70 percent of water withdrawals are attributed to irrigation (FAO, 1996). Table 1.1 illustrates the water use by sector and by continent.

Table 1.1 Water consumption by sector and by continent (after FAO, 1996)

Continent	Agriculture	Domestic	Industry	Total	
	%	%	%	km^3/year	m^3/cap/year
Africa	88	7	5	144	245
Asia	86	6	8	1531	519
Former USSR	65	7	28	358	1280
Europe	33	13	54	359	713
North and Central America	49	9	42	697	1816
Oceania (incl. Australia)	34	64	2	23	905
South America	59	19	23	133	478
World	69	8	23	3240	644

The demographic distribution of countries is rapidly changing with the urban areas growing faster - population growth plus urbanisation - than the agricultural population in all developing countries. Overall predictions for the year 2000 project that 50 percent of the global population will live in urban areas but rising to over 60 percent by 2025 (Serageldin, 1995). As a consequence, significant increases in demand for domestic and industrial water supplies can be expected.

This rapidly growing domestic and industrial demand for water will have to be met from reduced use in the agricultural sector. Projections of water demand of the International Food Policy Research Institute (IFPRI) indicate that global water withdrawals will increase by 35 percent between 1995 and 2020 with more than 80 percent of the increase being for industrial uses in developed countries (Pinstrup-Andersen *et al*, 1997). The share of domestic and industrial uses in developing countries is projected to double from 13 percent in 1995 to 27 percent in 2020. The more intensive use will also increase water quality degradation. Additional protective measures are required to safeguard the freshwater and soils from this increased urban pollution.

Competition among water uses already affects irrigated agriculture, especially in the water scarce regions in the semi-arid and arid zones. This pressure will become more severe and conflicts will increase as water supplies become more fully utilised. Solutions to these conflicts will require revisions of water law and systems of water property rights and water allocation. It is argued that in future, the price of water will be more closely related to its real economic value. This certainly will affect irrigated agriculture because the value of the marketable product per unit volume of water consumed is low. For example, only about 1 kg of maize grain can be produced in arid and semi-arid areas per cubic metre of water. For many semi-arid and arid areas, it will not be economical to import 1 ton of water for agriculture to produce 1 kg of grain even if renewable water was available. These areas will more likely be forced to use more local water for domestic purposes and import food[2] requiring increased expenditure in foreign currency.

1.1.4   Declining per Capita Arable Land

Soil is a basic, finite natural resource that is not renewable within a reasonable time frame. Only half the potentially arable land area of 3,200 million hectares is presently cultivated (Lal *et al*, 1988). Lal and Pierce (1991) indicate that the world per capita arable land area will progressively decline from 0.23 hectare by the year 2000 to 0.15 hectare by 2050, and to about 0.14 hectare by the time the world population stabilises in the year 2100. They also estimate that the minimum per capita arable land needed for an adequate diet is 0.5 ha under a modest level of inputs. The productivity of agriculture under rainfed, drained or irrigated conditions, must therefore increase to offset this decline in available area. Restoration of degraded lands is an ecological and socio-economic necessity to meet the basic necessities of the earth's inhabitants, including its human and animal populations (Lal and Stewart, 1992).

In the longer term, there can be no doubt that dramatically more food will need to be produced to meet the demands of a projected 8 billion mouths to be fed by 2020. A recent study by the International Water Management Institute (IWMI) (Seckler *et al*, 1998) concluded that that perhaps 80 percent of increase in food supply will have to be met from irrigated lands, due to production limitations in dry land areas in the developing world. Others, such as the late Ian Carruthers (1993) have suggested, controversially, that the bulk of the world's food supplies will eventually be grown in temperate Europe, North America and the Russian sub-continent.

1.2   DISAPPOINTING PERFORMANCE OF IRRIGATION AND DRAINAGE SYSTEMS

Despite the impressive gains in global food production during the last decades, policy and decision-makers in governments and in development institutions have raised criticism and concern on the performance and sustainability of irrigated agriculture.

The performance of many irrigation and drainage systems is significantly below their potential due to a number of shortcomings including:

---

[2]   By importing food crops no local water has to be allocated for its production. Hence, water is imported virtually. This concept is called virtual water.

+ poor initial design as a result of inadequate operational specifications, or design assumptions which were not or could not be fulfilled following construction and commissioning of the works;

+ distribution system layout did not adequately reflect existing land tenure or family/community associations in farm management;

+ poor, improper or inadequate management environments;

+ poor management systems within the managing organisations.

The most obvious manifestation of these shortcomings is unreliable main system water supply and poor maintenance practices. Improvements in on-farm water management are often hampered by the unreliability of water supply. Users are not motivated to organise themselves and participate in the operation and management of the water delivery system, and neither willing to pay water charges if the service is poor and unreliable. Furthermore, insufficient funding for maintenance and ineffective use of these funds results in problems of rapid deterioration of the irrigation and drainage infrastructure leading to further problems such as inequities in water distribution and reduced productivity. Combined with a lack of adequate drainage infrastructure, poor irrigation water management has been accompanied by soil degradation arising from waterlogging and salinisation of the irrigated cropland. To date, some 25 million hectares have become virtually unproductive as they are severely affected by these problems.

The earth summit in Rio de Janeiro (United Nation Conference on Environment and Development/UNCED, 1992) scrutinised the irrigation sector and its environmental impact and sustainability in that:

+ Irrigation uses too much water: about 70 percent of global fresh water use;

+ Irrigation leads to waterlogging and salinity: some estimates indicate that over 50 percent of the world's irrigated land has developed drainage problems, and some 24 percent is affected by yield reduction from salinisation;

+ Irrigation pollutes freshwater resources: through release of insecticides, pesticides and other agro-chemicals, and saline drainage effluents;

+ Irrigation affects human health: by creating habitats for vectors of water related diseases.

In addition to these issues, there is a widespread concern about the financial sustainability of irrigation. Investment costs for developing public irrigation schemes are almost always partly or fully subsidised and the recurrent cost of operation and maintenance is hardly ever fully recovered from the users. This means that irrigation often represents a permanent drain on government treasuries.

The rapid deterioration of the irrigation infrastructure worldwide is very severe. Most of the investment in irrigation development during the 1950s and 1960s was for development of new irrigated lands. Poor operation and maintenance practices have often resulted in a significant reduction in the economic life of the infrastructure and the concurrent loss of capacity to provide an effective water supply to farmers. Beginning in the 1970s until the present the emphasis of investment strategies progressively shifted towards rehabilitation and modernisation. Many irrigation schemes developed over the last decades have undergone periodic rehabilitation work as a result of premature decay of the infrastructure leading to the typical cycle of construction-deterioration-rehabilitation (construction). This recurrent need for rehabilitation of existing infrastructure confirms the lack of sustainability and other problems indicated above.

This cyclic infrastructure problem has been ascribed primarily to a lack of adequate funding for conducting appropriate maintenance. While this is the immediate cause of the problem, in fact it is the result of a lack of an appropriate and sustainable management framework. Such framework is needed to enable:

- The identification of the true cost of operating and maintaining an irrigation scheme;

- The implementation of appropriate policies to ensure that the level of revenues is sufficient to cover these costs.

The lack of sustainability has been one of the major factors in reducing the level of investment in new irrigation development. Moreover, future investments and development in irrigation are likely to be more difficult because of increased competition with other water uses and higher development costs for new projects as most low cost projects have already been developed. Worldwide, the irrigated area is projected to grow at an average annual rate of 0.6 percent per year during 1995 - 2020, less than half the annual growth rate of 1.5 percent during 1982 - 1993 (Pinstrup-Andersen *et al*, 1997).

The future challenges for the irrigation and drainage sector will be to develop and maintain a financially and environmentally sustainable operation that will enable it to meet:

- future food demand to keep pace with world population growth,

- increased competition for water between agricultural and non-agricultural users; and,

- a reduction in available agricultural land.

To meet these challenges, irrigation and drainage requires an integrated management of land and water resources that combine management of

- the water balance;

- the nutrient and salt balance;

- the financial balance.

1.3   GOVERNMENT POLICY AND PLANNING ENVIRONMENT

Until recently, substantial capital subsidies were justified in the name of food self-sufficiency and equitable distribution of benefits to subsistence farmers. It was often the intention of development agencies and national governments to obtain full payment of operational costs from the users. However, fee levels have consistently lagged behind the real costs of service and maintenance, and collections rates have varied from dismal (less than 10%) to poor (around 50%) in most countries. Faced with declining terms of trade in agriculture, reduced opportunities for new development and ever spiralling costs in subsidising the operation of existing systems, governments have been actively pursuing micro-economic reform in the irrigation and rural sectors.

Rural development, agricultural intensification and particularly irrigation development have faded from their accustomed priority in many national policies, as political pressure has risen to provide adequate water supply and sanitation to ever expanding cities. Despite greater opportunities for private capital investment in domestic water supply, the price-tag of such municipal services will require considerable injections of public and international loan money over the next 20 years (Serageldin, 1995; Briscoe, 1997).

The rising competition for water resources, either nationally or locally, where irrigation is already the dominant user is beginning to pre-occupy policy makers. Reallocation of "scarce" water resources and their use in higher value economic activities the world over now attracts global attention.

The broad policy setting that irrigation managers and planners now find themselves in can be summarised briefly as follows:

- increased competition for capital, leading to reduced investment, not only in new systems, but also for the deferred maintenance, rehabilitation and modernisation of old or under-performing infrastructure;

- a requirement by government that irrigation systems recover a higher proportion of recurrent costs of operation and maintenance from water charges;

- finance for improvement of systems should as far as possible, be provided by its users;

- improvement of the economic efficiency, particularly of large publicly owned and operated systems, through increased administrative efficiency, improved technical performance, reduced costs and increased gross output;

- more involvement of users in the management and assessment of system performance, partly as a quid-pro-quo for higher levels of fee payment.

Simultaneously, a number of developments are occurring within the agricultural sector at large:

- removal of production subsidies, particularly those on fertilisers, via promotion of increased private sector activity at all levels of the production and retail chain;

- reduction in the level and coverage of agricultural extension services and an increasing reliance on non-governmental organisations (NGO) activity in poverty alleviation and basic subsistence farming activities;

- simultaneous reduction in the level of funding for agricultural research and better targeting of research to more closely meet farmers needs;

- ever increasing concern with "stable" and "sustainable" farming systems;

- broad-scale approaches to land and water management (such as integrated catchment management), based on low-level community participation.

The dilemma for agricultural policy makers in developing countries is therefore how to attain an improved economic performance from existing irrigation and drainage infrastructure with minimal investment while requiring a longer term strategic plan for meeting future increases in food demand. To an extent both problems are congruent, in that the options for developing new irrigated lands are few and that productivity of land

and water will have to increase. Improving irrigation system performance now could be seen as a springboard for raising agricultural productivity later.

Broadly, there are six policy approaches to improve irrigation management and its economic viability:

+ reform of public sector bureaucracies from within, sometimes with redefinition of their primary roles - e.g. irrigation and drainage agencies becoming more involved with planning and management including regulation and less concerned with construction and infrastructure development;

+ transfer of responsibility for operation and maintenance, or transfer of ownership of part or complete irrigation systems to users - "Irrigation Management Transfer";

+ adoption of rational pricing structures that at least cover the cost of service of irrigation supply systems. This includes improved cost-recovery, preferably on the basis of volumetric measurement and charging for water deliveries;

+ use of economic instruments and market like mechanisms to help reallocate water to higher value uses - notably the authorisation of seasonal water trading and swaps within agriculture. In the longer term eventual allowance of permanent inter-sectoral transfers and the development of "water markets";

+ attraction of private capital in construction of new systems, or for ownership or management of complete systems as commercial enterprises;

+ sanction and often active encouragement of conjunctive use of groundwater to provide flexibility in water supply within rigidly operated surface irrigation systems.

Sometimes policy and social economic and legal organisation to implement it is not enough: policy makers tend to forget the strong interplay between technology and management (both social and economic). The design limitations of many water supply systems should not be overlooked and the opportunities to design for effective management when ageing or ineffective infrastructure is to be rehabilitated should not be missed. This in itself requires further training and capacity building and the inculcation of the philosophy of "design for management".

The irrigation industry not only must use less water and make transfers to other higher value uses, but also must make productivity improvements to grow more with less - more crop per drop or more dollars per drop-. Improved irrigation scheduling and water control will need to be supported by agronomic refinement and genetic improvement in the long term - items low on the policy agenda at the moment. In the medium term, it is hard to see how equity and subsistence considerations can take second place to economic efficiency in its purist sense, and successful policy will be one that finds balance on the tight-rope between self-financed sustainability and well targeted support for the rural poor.

1.4   LESSONS FROM THE PAST AND THE NEED FOR A NEW ORIENTATION TO IRRIGATION AND DRAINAGE MANAGEMENT

As discussed in previous sections, there can be no doubts about the pivotal role played by irrigated agriculture in keeping pace and at times surpassing the rate of increase in demand for food supplies. As well, it is clear that performance of many irrigation development projects has fallen short of producing the benefits envisaged during the planning stages. There has been ample speculation as to the reasons for these shortcomings. De Graaf & van den Toorn (1995) discussed some of the institutional aspects related to poor irrigation performance. They state that, "*What promised to be one of mankind's major steps towards intensification and expansion of agriculture has become an often disappointing investment, a burden on national governments and the focus of rent-seeking and tortuous communication between the farmers and agencies.* These authors identified the lack of reliable and responsive management of the main system as the main reason for poor performance claiming that instead of management of the system to deliver water in response to the requirements of irrigated crops, there is only system administration. The same authors provide a summary of the efforts of the irrigation community aimed at improving irrigation performance, concluding that each of the prescribed solutions at the time enjoyed that status of 'privileged solution[3]' that stifles the process of learning and improvement.

The concern and subsequent efforts to improve irrigation performance began in the 1970's and continues to these days. These efforts can be consolidated into five main categories:

- *Main System Development:* The decades of the 1950's until the 1970's were characterised by rapid development, dominated by massive investment in irrigation infrastructure heavily focused on the main civil works: dams and main distribution networks. Most of the development and subsequent operation of the infrastructure was driven by government agencies that placed a sole emphasis on construction to the detriment of operation and management. This role has often conferred them significant political clout, which together with their operation in an environment of absolute monopoly provided little incentive to be responsive and accountable to users;

- *On-Farm Development:* During the 1970's and 1980's substantial investment was directed towards the construction of infrastructure and implementation of extension programs aimed at promoting the adoption of better agricultural practices. These programs did little to improve the overall performance of irrigation mainly because the quality of the service provided by the irrigation agencies was not addressed;

- *Farmers' involvement:* Concurrently with a drive for on-farm development, substantial efforts were directed to increasing the participation of farmers in the various phases of irrigation development and management: planning, design, operation and maintenance. The emphasis was largely on the organisation of water users' associations as a vehicle for farmers to participate in the irrigation development and management process. By the end of the 1980's most irrigation agencies would claim to have in place some form of farmers' organisation, albeit many of them were only nominally organised. This approach with little or no possibility of changing the traditional top-down approach to development and management as usually imposed by the middle and higher level officials in irrigation agencies has produced little positive results as agencies failed to move towards more responsive and accountable management process;

---

[3] A privileged solution is not thought to require testing and modification. The answer will seem to lie at hand, and what matter is simply to find the resources and will to act (Morris, 1987)

♦ *Training of irrigation agency staff:* During the 1980's a great deal of emphasis was placed on training of irrigation agency staff as a form of institutional strengthening. Most of these programs however were ad-hoc at best as they were not designed to support a well-structured and focused management program for the organisations. In other words, while training has the capacity to improve the capacity of individuals to carry out their functions within the agency, the agencies themselves did not change from a construction orientation to a management focus;

♦ *Transfer of Irrigation Management:* During the 1990's the focus has shifted towards tackling the problem of management of irrigation by challenging the position and status of the irrigation agencies themselves in some form of the irrigation management transfer. Irrigation management transfer is a complex subject that has many forms, speeds and agendas in different parts of the world. The philosophy behind it is that users are more likely to operate systems effectively and according to their (often changing) requirements and also pay for the operation if they are also the system owners. Many farmers managed irrigation systems (FMIS) are small and have at some stage in their history been user financed, before government intervention and assistance in the latter part of this century. Withdrawing state finances in such cases is a straightforward case of cost shedding, although some countries make considerable investments in systems prior to "turnover" to farmer management, to ensure that they are in viable operating condition. Large surface irrigation systems are almost exclusively the product of substantial government investment and the idea of handing over systems of more than 100,000 ha to farmer management boards or even to franchise operators is daunting in the extreme. What sets large developing country surface systems apart from those in say Australia or the USA is the enormous number of users who are supplied. As a result, management transfer at this scale envisages user ownership or operation at the tertiary and secondary level, with the main system and head works operation being undertaken by the (state-owned) irrigation agency. Nevertheless, opportunities for increased public participation in the management of very large systems exist in the form of Management Boards and other umbrella bodies for water user associations and tertiary block "companies".

Despite these changes in development and management emphasis, continuous and increasing concern developed over the low performance of irrigation, especially in less developed countries. It is argued in this book that regardless of the nature of ownership of the irrigation and drainage organisation and its assets, the type and characteristics of services to be provided and the environment in which the organisation must function remain unaltered. These are water supply for irrigation and removal of excess ground- or surface water and salts in an environment of natural monopoly. Furthermore, it is clear that a common shortcoming that is observed throughout the evolution of irrigation development this century is the lack of service orientation in the management and operation of irrigation systems. This aspect is obviously more critical in large systems than in small systems given the complexity involved in the operation of the extensive canal irrigation networks. Newell (1916) who drew attention to the issue of quality management in the early stages of irrigation development in the Western United States identified the criticality of system operation and management. In his book on "Irrigation Management" he stated: "*Planning and building is only the beginning; the really difficult and at times discouraging work is that of properly utilising the irrigation systems after they are built and of getting fair returns from the irrigated lands".* Newell's book contains a detailed description and analysis of the difficulties facing the managers of large-scale irrigation schemes. Building on Newell's work, Snellen (1997) summarises the key concerns and insights about the sustainability of irrigated agriculture (Box 1.1).

This monograph builds on these observations and other lessons learned. The main lesson is that the focus of irrigation development should be on the development of building of a management capacity that is based on the mutual interest of irrigation authorities, and water users. In the past irrigation authorities always managed to stay out of the "line of fire" as the primary focus was on construction and later on the role of farmers. The role of the irrigation authorities, though critical to achieving high performance, has seldom been disputed.

It is recognised that the development of high quality irrigation management requires a

---

Box 1.1 Early lessons learned from early irrigation development (Snellen, 1997)

- Capacity building for sustainable management of an irrigation scheme is more difficult than the design and construction of the scheme;

- Capacity building for sustainable management is more difficult in a government-developed scheme than in a scheme constructed by a farming community;

- The potential for accelerated irrigation development through government-constructed irrigation schemes is limited by the human and financial resources available for building the capacity for sustainable management;

- It is essential that Governments are aware that the capital invested in a new public irrigation scheme will yield an economic rate of return that is lower than the opportunity cost of capital.

---

competent authority that is accepted by the users and that this can only occur if it responds to their needs and ambitions. In this monograph the road to professional management is explored with effective accountability mechanisms after a period of – often forced – establishment of water user associations. On the one hand, users will control the management authority for quantity and quality of service while on the other hand the government will exercise control for effective and efficient use of public resources. The purpose of this monograph is to provide the reader with the concept of service oriented management focusing on the qualities of the irrigation organisation that are required to achieve this service orientation and the processes that must be implemented to achieve these qualities.

Chapter 2

The Management Environment of Irrigation and Drainage

One of the difficult problems facing water resource managers is to integrate viewpoints into workable strategies, and the starting place must be to understand the viewpoints

N. Grigg, 1996

Irrigation and drainage authorities deal with the provision of services and transactions related to water which, as an economic good has a different character for each service. This difference in character requires different approaches to management and customer relations. Irrigation water can be regarded as a common pool resource. Its management requires a clear set of rules for its use and appropriate mechanisms to penalise those who transgress them. These rules must be agreed and accepted by all the players involved. Where tight control on the distribution can be enforced, water can also be regarded as a private good. Irrigation and drainage infrastructure can be regarded as a public good, especially in large irrigation systems. The players in irrigation differ from those involved in drainage in that the provision of drainage services is not exclusive for farmers but may also involve non-agricultural land such as urban and industrial areas.

2.1 COMPETITION FOR WATER

Irrigation and drainage are generally concerned with the use of land and water resources for agriculture. These resources are finite and often scarce and must be shared with many other individuals and organisations for many different purposes. This is not only the case for water use within the irrigation scheme but also between the irrigation and other water uses at the (sub-)basin level. Policies and actions to use or control water for these different purposes are aimed at improving security, social and economic well being and the preservation of the ecosystem. These water use and control activities which exist between different water uses and within a particular use such as irrigation may conflict with each other and have the potential to create various kinds of problems. Such problems are referred to as externalities, open access, public interest and scarcity problems (Lord and Israel, 1996).

Externality problems exist when actions of one party affect the well being of a second party, and the first party does not gain by considering this effect and reducing the resulting impact. For instance, an industry that discharges polluted effluents upstream of the intake of an irrigation scheme. The polluted water will affect agricultural production but the industry will not gain by building a wastewater treatment plant.

Open access problems exist when the use of the resource is open to all, and when the rate of use of that resource affects the amount that can be used. For example, a river basin in which the amount of water is insufficient to supply all users and uses in the desired amount. Without proper allocation rules some uses and users will take what they need at the expense of others.

Public interest problems relate to the need for providing a particular good to all in equal amounts. No one can be excluded from consuming it, and the cost of providing it to one is the same as the cost of providing it to all. These goods are likely to be under provided because no one will undertake to produce them since they cannot be withheld from others and cannot be sold at a profit. Main drainage and flood protection works are good examples of this type of goods, which are normally provided by the government.

Scarcity problems exist when the users desire more of a good than the quantity available at a given price. Economic markets handle scarcity by allowing competition. Those with a greater purchasing power and to whom the resource is most valuable will outbid others. Public institutions such as river basin authorities or government will need to put in place adequate mechanisms to safeguard the low-income segments of society and the needs of the environment from the negative effects of scarcity.

2.2 CHARACTERISTICS OF WATER AND RELATED SERVICES

The problems described in the previous sections arise because of the physical attributes of a resource. These attributes affect the relationship between users and potential users of the resource. Two independent attributes - *feasibility of exclusion* and *subtractibility* - are used to classify goods (Ostrom and Ostrom, 1977). Exclusion occurs when potential users can be denied access to goods unless they meet certain criteria. A good is said to be subtractive when one person's use of the good precludes its use by others.

If these two attributes are arranged in a simple matrix as illustrated in Table 2.1, the combination of cost of exclusion and subtractive consumption results in four logical types of goods: Private goods, Public Goods, Toll Goods and Common Pool Resources (Ostrom and Ostrom, 1977). Common Pool Resources are also referred to as Open Access Goods (The World Bank, 1993).

Another important attribute of goods and services is related to their economic value. Although water supply and drainage services can be generally regarded as a business in the commercial sense, water itself has some special characteristics:

   ♦   water is a finite renewable resource and as such it can be considered as a scarce good. It cannot be "produced" or "sold" indefinitely or for as long as there is a demand. Irrigation water has to be shared with other users and has a strategic value to maintain quality of life and environment. It therefore cannot be considered as a resource that can be allocated solely on the basis of market forces or purchased by the highest bidder. The use of the natural resource water must therefore be constrained by a set of regulations to protect these strategic values of water for society;

• water supply services are normally provided in a situation of monopoly. Usually high capital investments are needed to provide these services and the levels of turnover from the provision of service in relation to capital investment are generally low. This makes competition between service providers impractical, making necessary a separate set of regulations for consumer protection.

Table 2.1. Logical types of goods according to their subtractibility and cost of exclusion

	Subtractive Consumption	*Non-Subtractive Consumption*
Costly to exclude	Common Pool Resource (e.g. Irrigation, Water Resources, Fishing Grounds)	Public Good (e.g. Drainage, Flood Control)
Not Costly to exclude	Private Good (e.g. Bread)	Toll Good (e.g. Cable TV)

Market forces generate an efficient (economic) allocation of resources when the market is competitive. However, economic activities in which there are economies of scale (a large fixed cost relative to variable cost) or economies of scope (lower unit cost of producing several products in combination rather than separately) tend to become "natural" monopolies. In such a situation, a single supplier dominates the market. In a market monopoly, resource allocation is inefficient since monopolies tend to produce less and charge more for any good or service than under competitive market conditions. Furthermore, because the threat of entry of potential competitors is minimal, incentives for innovation and dynamic efficiency are diminished (The World Bank, 1993).

Overcoming potential conflicts between different water uses requires creating an effective set of rules and regulations for managing the water resources within an adequate policy and legal framework. The absence of such a framework precludes planners and managers from establishing effective accountability and regulatory mechanisms to address these conflicts when they arise. Conflicts may also arise between users within a particular water sub-sector such as irrigation, which must be dealt within the respective sub-sector.

2.2.1 Institutional Arrangements for Managing Common Pool Resources

In reference to the logic of collective action, Olson (1965) states that self-interested individuals will not act to achieve their common or group interest unless the number of individuals is quite small, or unless there is coercion or some other special device to make them act in their common interest. An individual who cannot be excluded from obtaining the benefits of a collective good once the good is produced has little incentive to contribute voluntarily to the provision of that good. Moreover, Hardin (1968) argues that the "Tragedy of the Commons[4]" is unavoidable unless "the State" backed by coercive powers controls these commons to prevent their destruction. Others recommend that privatising those resources will resolve the problem. However neither the state nor the market is uniformly successful in enabling individuals to sustain long-term, productive use of natural resource systems. Empirical evidence shows that without effective institutions, common pool resources will be "under-provided " and "overused" (Ostrom, 1993). This seems to be an inevitable consequence in those situations where the participants have no involvement in the process that determines the situation they face (Ostrom, 1986). In situations where an identifiable group of individuals is involved and everyone is aware of the effect of their actions on others, individuals can develop a set of institutional arrangements that changes the structure of the situation they face. A set of institutional arrangements that can effectively monitor and impose sanctions on those who breach the rules creates incentives for individuals to co-operate in water allocation, distribution and investment.

Another option to avoid the tragedy of the commons is co-operation between interest groups. This enables the solution of three institutional problems: (i) consensus about the organisational structure, (ii) commitment of interest groups, and (iii) participation of individuals in mutual control of the compliance of agreements, regulations and rules made to govern the commons.

2.2.2 Typology of Water Related Services

The type of good is important in defining the delivery of services and the relation between the service provider and its clients. Tang (1994) characterises irrigation systems as common pool resources because of two main properties:

- the flow of irrigation water available at any time in an irrigation system is limited. The withdrawal of water by one individual subtracts the same amount of water available to others;

- once an irrigation system is constructed, it is costly but not necessarily impossible to exclude potential beneficiaries from using the system. Unless institutions change the incentives facing appropriators, one can expect substantial over-appropriation of water and under spending for operating and maintaining the infrastructure.

Therefore, operation and management of irrigation systems requires coordination among many participants sometimes with diverging interests. While individual participants attempt to advance their own interest, their actions sometimes produce outcomes that benefit themselves at the expense of others resulting in unintended and harmful consequences for themselves as well as for others. An example of such behaviour is the

---

[4.] The tragedy of the commons arises when many individuals are involved and have difficulty in communicating and enforcing agreements among themselves leading to the destruction of the commons through over-appropriation.

over-appropriation of irrigation water at the head-end of an irrigation system causing water stress or scarcity for tail-end users. As a result, tail-end users refuse to participate in operation and maintenance activities and cost sharing leading to the system falling in disrepair. An understanding of these various situations faced by the participants and the resulting incentives for certain behaviour can assist with the design of rules to be used in irrigation system management.

The view of irrigation as a common pool resource is valid for small-scale farmer managed irrigation systems. However, in large-scale schemes, the basic interpretation of exclusion and subtraction requires more elaboration. Systems where the main works (dams and main canals) are owned and operated by a government organisation can also be regarded as public goods. If on the other hand, exclusion of users can be arranged, the water delivered can also be regarded as a private good.

Drainage systems are different in character. Drainage systems are built to allow individuals to dispose of their excess water and salts[2]. Sometimes drainage systems are also built to control surface and ground water levels. Once a drainage system is constructed, it is costly to exclude individuals and organisations to make use of the system. Moreover, the use of the system by one individual does not preclude its use by others. Under these premises, drainage systems can be classified as public goods. The operational use of public goods does not need co-ordination between its users because of the non-subtractability. However, the maintenance of the system and the possible negative environmental impacts require collective action. From this point of view, the public good can be treated the same as the common pool resource.

In summary, drainage and flood protection systems can be considered as public goods because of their non-subtractive consumption and low level of exclusion. Proper management of both goods require collective actions like maintaining facilities and management of water that allow all individuals participating in its use to obtain maximum benefit.

2.3   IRRIGATION AND INTEGRATED WATER RESOURCES MANAGEMENT

Irrigation and drainage services operate and interact with other water users within the water resource base as a consumptive user (through evapotranspiration) and as a polluter (salts and agro-chemicals in the drainage effluents). While irrigation itself can be considered as a common pool, it also operates within the larger common pool of water resources. Water consumed by irrigation is not available for other purposes (subtractive consumption) and it is often difficult to prevent irrigation systems from taking water from the river or groundwater reservoir unless clear rules and control mechanisms are in place. This set of rules and control mechanisms to manage water within a hydrologic unit for different purposes is termed Integrated Water Resources Management (IWRM)

[5.]   Often drainage systems are used as a source for water especially in irrigation areas where irrigation runoff is reused. An over-appropriation at the head-end leads to wastage returned to the drainage system. This water can be reused further downstream. In some instances, diversion works are built in the drain, often creating a disturbance of the drainage function. Howver, very few drainage systems provide individuals with any control especially in developing countries.

2.3.1 Aim of IWRM

The management of water withdrawal for a particular use requires that the potential impact on the quantity and quality of the resource available to other uses is properly assessed. IWRM is a rational approach to manage the competitive use of water and its potential impacts by taking a holistic view of the water resource system. It presupposes a change from the often-narrow management perspective of one sub-sector by one government agency to strive for a participatory multi-sectoral management perspective that involves all the stakeholders. IWRM therefore integrates all natural aspects of water resources, all sectoral interests and stakeholders, the spatial and temporal variation of resources and demands, relevant policy frameworks and all institutional levels.

Verhallen *et al* (1997) define IWRM as "... *the management of surface and subsurface water in qualitative, quantitative and ecological sense from a multi-disciplinary perspective and focussed on the needs and requirements of society at large"* In practice this implies a recognition of the fact that:

♦ sustainable use of water by humans, flora and fauna requires ecologically healthy functioning water systems;

♦ in managing these systems, all interests need to be taken into account and that regulation is required to guarantee its sustainable use.

♦ Stakeholders interests can be best represented by establishing a co-ordinating body with planning, co-ordination, decision making and policing powers.

2.3.2 Functions and Functional Levels of IWRM

A number of factors need to be assessed in the implementation of IWRM (IDB, 1997) which relate to the context in which policy is pursued and programs are developed, and to the decision making and co-ordination process. These processes are formulated and carried out through three main functions, namely:

♦ operational or water use function;

♦ organisational or water resource management function;

♦ constitutional or water policy and law function.

The *operational function* focuses on the use or control of water for specific purposes to fulfil specific needs and demands. These management activities are carried out by first line agencies (fig. 2.1). These may include irrigation and drainage, water supply and sanitation, flood protection, hydropower, industrial supplies, recreation, fisheries, navigation and the preservation or rehabilitation of ecosystems.

To minimise problems arising from conflicting interests between the different uses and users, co-ordination of water use and water allocation is required. Addressing these problems often also requires establishing new rules or changing the existing rules for water use. This is the *organisational function*. It involves co-ordination, planning, decision making and application of water use rules in water systems (river basins, aquifers). Often a river basin authority is assigned to execute this function.

An appropriate enabling environment must be created to make the organisational function possible. The *constitutional function* of IWRM involves the formulation of water policies, institutional development policies including human resources development and normative and executive legislation.

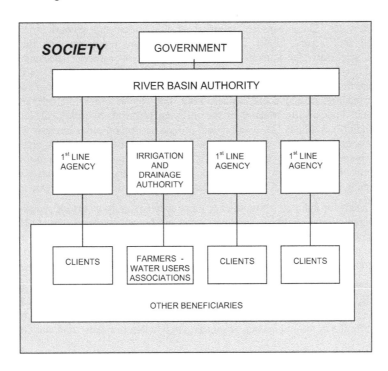

Figure 2.1 Irrigation and Drainage in IWRM Context

2.4   IRRIGATION AND DRAINAGE IN THE CONTEXT OF IWRM

As discussed earlier, the IWRM process aims at harmonising the desires of society with regard to alternative uses of water and the functioning of water systems through technical and legal interventions. In principle, the boundaries of the system are determined by the objectives of IWRM, however, the natural hydrologic boundaries of water system are commonly adopted for this purpose.

The needs and demand for water by society and the desire of government to reconcile these with the actual capacity of the water system determine the management priorities for the system. Systems can then be managed to meet the water demand and quality norms or standards for each use and maintain or enhance the overall environmental quality of the system. The imposition of such norms would imply for instance that irrigation and especially drainage effluents would be subjected to a set of water quantity and quality criteria based on the water quality required for irrigation and water quality tolerance imposed on drainage effluents to meet environmental standards.

Conflicting Interests

There are many parties with conflicting interests within a given water system. Individuals, groups and nations have interests in obtaining, preserving or enhancing their use and control of water. These can be classified as first order and second order interests (Hofwegen and Jaspers, 1999). First order interests are represented by the essential conditions necessary for human and aquatic life in the system. The importance is confirmed in the Dublin and Rio the Janeiro Conferences in 1992 (box 2-1). Second order interests are those which can, if necessary, be prioritised after being weighed according to economic, ecological and social values. First and second order interests are system-specific and time-specific because of the different physical, hydrological, cultural and socio-economic conditions. As development progresses second order interests will change In water resources allocation there is general agreement that first priority is given to water for basic human needs. Second priority is usually assigned to water requirements to maintain essential life support systems or ecosystems. Equity principles play a major role in establishing these priorities which are considered part of first order interests.

All other needs for industry, agriculture or other society should be prioritised according to socio-economic criteria which consider water to be an economic good. Here, it is important to note that although cost-recovery and economic pricing are over-riding principles, pricing and tariff regulations within sub-sectors are often considered necessary where equity or social well-being are at risk or the environment may be endangered.

In many water resource systems, the irrigation sector will be in competition for water with other users. In this competition, the socio-economic importance of irrigated agriculture will be judged by society against the benefits of other uses in three main areas: (i) whether it responds to the needs and desires of society, (ii) whether it makes good and efficient use

---

Box 2.1 Water and Environment: The Dublin and Rio de Janeiro Conferences in 1992

The International conference on Water and Environment: Development Issues for the 21st Century, held in Dublin, called for new approaches to the assessment, development and management of freshwater resources. The conference report sets out recommendations for action at the local, national and international levels, based on four principles:

♦ The effective management of water resources demands a holistic approach linking social and economic development with protection of natural ecosystems;

♦ Water development and management should be based on a participatory approach involving users, planners and policy makers at all levels;

♦ Women play a central part in providing, managing and safeguarding water;

♦ Water has an economic value in all its competing uses and should be recognised as an economic good

The United Nations Conference on Environment and Development in Rio de Janeiro confirmed the importance of the holistic approach. It was stated that : Integrated water resources management is based on the perception of water as an integral part of the ecosystem, a natural resource and social and economic good.

of resources and (iii) whether its impact on the environment is within acceptable limits.

2.5   MAIN INTERESTS AND INTEREST GROUPS

Systems of irrigated agriculture have been established to achieve objectives such us to enable food and fibre to be produced or to assist in improving the social and economic well being of farmers. Obviously, farmers are in a central role in the activities necessary to meet these objectives. However, three groups with vital and primary interests in the successful functioning of an irrigation and drainage system must be considered, each of which may have a different perspective of the role of agriculture in the system and its management priorities. They are the *farmers* (use level), the *authority* or agency responsible for the management of the irrigation and drainage system (operational level), and the *Government* representing the interest of the country and society at large (constitutional level). A formal or informal body can perform the co-ordinating role with other water sub-sectors (the organisational function). Each of these groups may have different objectives and expectations as to the outcome of their efforts.

Farmers

The objectives of the water users are largely of micro-economic nature. The primary concern of farmers is to produce enough food to maximise their family income or for subsistence, and in the longer term to move from a position of subsistence farming to a position of "farming for profit". The farmer's production is affected by a number of physical and environmental factors including soil, climate, water supply, pest and diseases; and other non-physical factors including availability of labour and capital, land tenure, financial support, markets, culture and tradition.

Farmers have a vital interest in the operations of the irrigation and drainage authority. Assuming other conditions are met, the productivity of the farmer's enterprise will be determined by how well irrigation water is supplied, drainage water is discharged and groundwater levels are maintained.

To obtain maximum yields, farmers require flexibility in water supply in terms of frequency (how often), rate (how much) and duration (how long). The ideal mode of water delivery from a farmer's viewpoint is delivery at will or "on-demand". In this way, (s)he can match irrigation scheduling to soil and crop needs and to rainfall. Moreover, (s)he can coordinate irrigation with other farming practices. The concern of most farmers in relation to irrigation is the reliability and predictability of the water supply.

Regarding drainage the primary concern of farmers is the prevention of inundation, too high groundwater tables or waterlogging and the often consequent salinisation (Schultz, 1990).

Irrigation and Drainage Authority

The organisation managing an irrigation and drainage scheme can have different organisational configurations. Regardless of what type of organisation is in place, it has a number of objectives that it must attempt to meet. The main task of the authority is to provide an adequate water delivery and water removal service to the various users. In doing so, it will attempt to achieve optimal productivity at the farm level and at the scheme level within the constraints imposed by the system infrastructure, the water development policy and the changing environment surrounding water management and agriculture.

To achieve optimal productivity at scheme level, rules are required on how water is to be distributed, especially in time of shortages. The same can be said for drainage in case of high drainage loads where sometimes temporary inundation has to be accepted. Often the rules to apply in these cases are set out in the development policy established during the planning phase of the system. These rules can also change if the conditions of agriculture such as cropping pattern change. In some cases, the objectives of the authority may in fact be in conflict with those of the individual users.

Frequently, the organisation also faces constraints imposed by Government policy to ensure certain minimal conditions of water supply service to farmers or prevention of environmental damage. Taking into consideration all the constraints imposed by the environment surrounding irrigated agriculture, the organisation in conjunction with users must formulate clear rules for operations and water delivery and drainage that will be embodied in a set of *level of service* specifications.

Government

The Government has a number of economic, social and environmental objectives for the country, which the irrigation system and/or drainage systems are supposed to fulfil. These could include amongst others increased food production, employment generation, generation of foreign exchange or alleviation of poverty. Some of these objectives will involve direct interactions between the Government and the irrigation and drainage authority. Some will involve the Government and farmers and other stakeholders. Especially involvement is required on issues that will include levels of subsidy in the establishment period of a project, marketing and prices for agricultural production, credit and financing, cost recovery from farmers as prosperity increases, and financing of the operation of the authority.

There is the realisation that the river basin is the logical unit for water resources management. Although this is undoubtedly true, it must be borne in mind that this management framework must function within the overall objectives and policies of Government. River basin management is largely an operational matter that involves water allocation, water quality management and stakeholder involvement and while river basin co-ordinating bodies may be responsible for policy making and for setting objectives and strategies for operational management of the basin, the ultimate legislative function still lies with government.

Government has several roles to play in relation to irrigation and drainage. Firstly the government utilises irrigation and drainage development and management as a tool for social and economic development. Secondly the government must ensure that the negative effects of irrigation and drainage do not harm the interest of society. Thirdly, the government must create mechanisms to allow an effective operation or irrigation and drainage within the overall water resource system. As a result, the government has to create an enabling environment for irrigation and drainage authorities to function both as a service provider and as one of the water users.

2.6   TYPES OF IRRIGATION AND DRAINAGE ORGANISATIONS

The provision of irrigation and drainage services entails a set of functions that the organisation must carry out. The organisation in charge of providing this service needs to formulate clearly stated objectives, goals and strategies for its provision. To provide this service, the organisation must operate assets on behalf of its owners or shareholders. It also uses resources and spends and collects money as payment for water service. Irrigation organisations are in most cases a retailer of water although it can also be a bulk supplier. Clearly, such organisations have all the characteristics of a business organisation established for the provision of water supply or drainage services.

This set of complicated and interrelated functions must be managed properly to ensure that the service reflects the needs and aspirations of its clients and is provided in the most effective manner.

There are a number of organisational scenarios that apply to irrigation and drainage organisations around the world. Many irrigation organisations are still funded from central treasuries and this situation to some extent "shields" the organisation from the normal commercial pressures. Increasingly, governments are opting for either a partial or full disengagement from the management of irrigation by transferring responsibility to users. In this context, it is useful to provide a classification of organisations and the nature of the legal and financial links that develop with the users and the Government (Lee *et.al*, 1995). Figure 2.2 illustrates this classification. The "agency or organisation" is an individual or group of individuals providing services, with the Government having no direct involvement in the provision of the service but with the capacity to provide the legal framework that regulates the provision of service and land and water rights. It should be understood from this definition that the "government" function includes the role of co-ordinator between the different uses on (sub-) river basin level. A private agency can provide a service to users under a contractual relation. In this case, the Government has no intervention in the contract but may set the regulatory framework for it. A private agency is characterised by a full cost recovery operation, which distinguishes it from the other types of arrangements.

The organisation can be controlled by Government and supply a service to users. In this case, the Government may also control the use of resources that are provided to the organisation to provide the service. In all these types of arrangements the provision of service may be carried out under formal "contracts" or informal "customer agreements". Formal contracts usually are made between the organisation and its customers and must be consistent with the limitations imposed by Government regulators. Disputes between service provider and customers are normally resolved in an informal basis, or through the normal legal channels. Customer agreements normally rely on service targets established by mutual consultation between the organisation and customers. The quality of service provision is usually monitored against these targets and adjustments are made within the framework of the operational process.

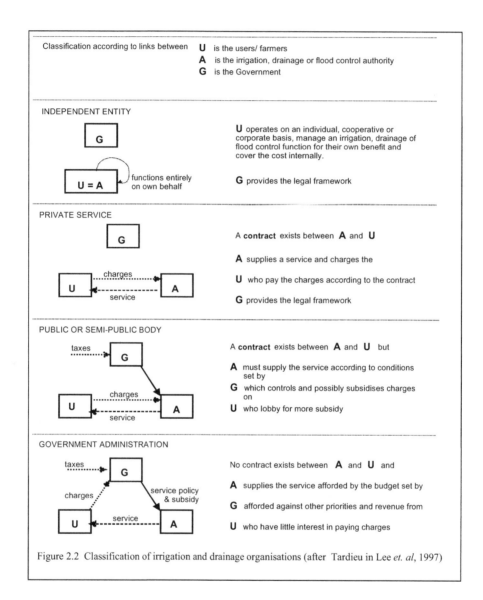

Figure 2.2 Classification of irrigation and drainage organisations (after Tardieu in Lee *et. al*, 1997)

2.7   SERVICE ORIENTED MANAGEMENT: RATIONALE AND KEY ELEMENTS

An irrigation system enables water to be acquired, transported and supplied to users. A drainage system allows (ground)water level control and/or excess water, salts and other pollutants to be removed from the cropland and other non-agricultural areas. To carry out these functions, an hydraulic infrastructure that consists of facilities for diversion, conveyance, regulation and measurement of water and an organisation in charge of its management is needed. In discharging its functions, an irrigation organisation must meet several accountability criteria including liability associated with performance of its functions, political and social responsibility embodied in the effectiveness of the organisation in meeting the expectations of the Government, and the farmers (clients). To meet these diverging interests, irrigation and drainage organisations are established with the purpose of managing the business of irrigation and drainage in a financially and environmentally sustainable manner. The irrigation and drainage authority must also comply with water allocation and discharge decisions made by the river basin co-ordinating body as part of the overall water management policy at the (sub-)basin level.

The nature of irrigation and drainage systems as both Common Pool Resource and Public Good was discussed earlier in this chapter. Then it was argued that the provision of both irrigation and drainage services occurs largely under conditions of monopoly. It is therefore obvious that some form of consensus is required among the irrigation and drainage organisation, the Government, other affected parties, and if present the river basin authority, about the type and form of service to be provided by the irrigation and drainage organisation. This consensus must be achieved on the basis of the associated cost of meeting the agreed service standard, the desire of users to obtain this service, their willingness to pay for it, and the existence of an appropriate legal framework that will ensure the enforcement of rights and obligations of the participants.

The overall success of the irrigation and drainage system in general and the authority in particular is measured both in terms of the effectiveness of its internal management processes, its effectiveness to contribute to the achievement of the overall Government's social and economic objectives, and in meeting the expectations of farmers.

The ability of the irrigation and drainage organisation to respond to the needs of its customers reflects its level of "service orientation." (Box 2.2). Output orientation is a key feature of service oriented management in that the actual cost of service provision is inextricably linked to the provision of service. This ensures that users become fully aware of the cost associated with the provision of the service. They can make informed decisions on the level of service they are prepared and willing to pay for. This interactive process is critical to ensure the necessary commitment from the interested parties. Moreover it should to facilitate the balancing of appropriate "trade-offs" between improvements to the main system in timing and extent, and options for alternative investment and improvement at the farm level to achieve desired levels of flexibility in water delivery systems.

The ability to enforce the agreed service standards is crucial to ensure its delivery and compliance with its specifications and conditions by the parties involved. This can take various forms including service agreements with legal contractual status between service provider and recipient. These agreements give a detailed description of services to be provided, payment in return for services, monitoring and verification of service provision, consequences of failure to comply with agreements by both parties and rules for arbitration of conflict.

In summary, effective irrigation and drainage operations, especially within the context of IWRM require a management system that responds both to the needs of the farmers and the needs of society at large. The irrigation and drainage authority will find itself on the one hand as a service provider to farmers or farmer groups and at the same time as a client of the authority managing water resources on a (sub-) basin scale. For the irrigation and drainage organisations to carry out its service provider role effectively it must do so within clear rules for quantitative and qualitative water abstraction and discharge of drainage water established by the river basins authority.

Box 2.1 Key elements of sustainable Service Oriented Irrigation and Drainage management

- It is output oriented: the cost of service provision is based on well developed operation and asset management programs

- It involves users to determine levels of service and the associated cost of service

- The irrigation and drainage organisation should be able to recover the cost of service provision either from direct consumer payment or from subsidies

- It relies on an appropriate legal framework that provides protection for users, the organisation providing service and the general interest of society

---

Chapter 3

Management Processes of the Organisation

Planning is designing the future and then inventing ways of bringing this about

R. Ackoff, 1981

Effective service orientation in irrigation and drainage management requires a significant shift in the institutional and management capacity of the irrigation and drainage organisation. Organisations must be able to formulate and implement strategic (corporate) plans with clear objectives and strategies to achieve them. These must be fully integrated with adequate financial planning across the organisation. Central to the success of strategic planning is the ability to assess progress towards achievement of goals set out in the strategic plan. Performance measure must be viewed as an integral and on-going activity within the management process. Key performance indicators must be developed to measure organisational and management performance consistent with the goals and objectives of the management plan.

3.1 IRRIGATION AND DRAINAGE ORGANISATIONS

Irrigation and drainage are activities that support agricultural production by supplying water to supplement natural rainfall, controlling surface and groundwater levels and removing excess water from the cropland. As such, they can both be considered services to agriculture, which are provided in most cases through an organisation being an irrigation and drainage authority. Single user and farmer managed systems are often associated with small-scale systems and are not the focus of this monograph.

In Chapter 2 the management environment or the external processes to irrigation and drainage authorities was discussed. It was shown that there are three main protagonists with differing roles which participate in this process, namely: The Government (including the river basin authority), the irrigation and drainage authority and the farmers. The focus of this chapter is the internal management process required for the irrigation and drainage authority to provide effective irrigation and drainage services. The authority in the context of this study describes any type of organisation that is in charge of providing the irrigation and drainage service. These may include Government operated agencies, private agencies or individual water users associations.

3.1.1 Defining Organisations

Irrigation and drainage services involve a number of tasks in managing an irrigation system from head works to farms and a drainage system from the cropland and other surrounding areas to the outfall. These activities with their generally large geographical spread require the involvement of a number of people with a variety of skills in an organised manner.

The study of organisational theory shows a number of definitions of organisations from various authors. Gerken (1995) defines organisations as "...*Social entities constructed by deliberate human design that serve the specific purpose of the organiser.*" This aspect of the definition often leads to the undesirable situation of organisations becoming an end in themselves. Daft (1995) defines organisations as "*Social entities that are goal-directed, deliberately structured activity systems with a permeable boundary.*" This definition departs from Gerken's definition in that organisations are conceived to have permeable boundaries. He further elaborates on the elements of this definition stating that social entities are composed of people or groups of people, which together with the roles they perform, are the building blocks of organisations. Furthermore, organisations are goal directed if they exist for a purpose. Therefore, organisations perform tasks that are deliberately subdivided into sets of activities that require to be performed in a structured manner. Robey's (1986) definition of organisations is more comprehensive and better encapsulates these roles and functions of organisations. He defines organisations as "...*A system of roles and stream of activities designed to accomplish shared purposes.*" This definition presupposes:

- a systems of roles describing the structure of the organisation,
- a stream of activities referring to organisational processes,
- that they are designed by human actors; and
- they exist because of shared purposes.

All organisations have boundaries that separate them from other organisations. These boundaries however should be permeable in order to share information and technology. This is especially the case between an irrigation and drainage organisation and other water and agricultural organisations where cross interactions between them are necessary to co-ordinate policy and management decisions aimed at maximising farmer productivity and use of water.

3.1.2 Deficiencies of Irrigation and Drainage Organisations

The development focus in the period between the 1950s and 1980s has oriented the organisational structure of many irrigation and drainage agencies to a development purpose rather than to a management purpose. This has often led to the neglect of irrigation and drainage authorities of post-commissioning management of the irrigation infrastructure. This has resulted in lack of responsiveness to the requirements of irrigated agriculture as the development focus was usually combined with a lack of accountability to the users. Often, this is further compounded by the dynamic nature of irrigated agriculture derived from the continuous market changes and competing interests with other water resource users. The main feature of this type of management is the lack of clear goals and objectives related to the purpose of irrigation and drainage and the provision of services. A budgetary process whereby budget needs and commitments are not related to the achievement of service specific goals often accompanies this.

Having in most cases become large and single providers of specific services, the irrigation and drainage authorities have mistaken the nature of the scale advantages. Their prevalent view has become one of seeing customers as passive and unproductive rather than active and productive, failing to convey management information to the consumer, improve the quality and productivity of human resources and promote the adaptation of services to the needs of clients. This all can be brought back to the fact that accountability in the system was lacking and that users have had no or very little say in how systems are managed. This was often justified by the fact that they did not pay or paid very little for the services.

Another important factor is that the members of the organisation have not or only partially supported the purpose of the organisation. Meagre career opportunities arising from poorly developed human resources development programs and poor remuneration frequently lead staff to low morale and motivation. It also encourages members of the organisations to look for opportunities within or outside the organisation to improve their personal situation at the cost of service provision.

An important element arising from all the definitions of organisations is that organisations are not an end in themselves. This is especially significant in the irrigation and drainage sector where many organisations exist for themselves; they tend to be more concerned with inputs and internal processes rather than focussing on outcomes. This is often the case because they are funded from central Treasuries, lacking a commercial orientation aimed at providing the most cost-effective service.

3.1.3   Service Oriented Irrigation and Drainage Organisations

Service oriented management is heavily output oriented and is designed to meet the provision of irrigation and drainage service according to a clearly defined *level of service*. This level of service becomes the key set of goals and objectives for the organisation to define its activities and budgets and against which the performance of service delivery must be measured.

To achieve its objectives, the irrigation and drainage organisation needs to formulate and setup adequate management systems and strategies. The science of management has dealt with this type of problems in the corporate industrial sector for many decades. Despite the public sector having been traditionally a service sector, the corporate management concept has begun to be applied only recently (Victoria State Government, Australia, 1986). The implementation of fully integrated corporate planning to public sector services implies big changes to the way the organisation conducts its business. Despite all the complexities and peculiarities in the administration of large scale government managed systems, there is no apparent reason that the principles of service management as discussed in Chapter 2 should not apply equally to private and to public organisations.

The key to this process is for the public organisation to become consumer and service oriented. This requires a significant shift in the management capacity and attitude of the irrigation and drainage organisation. Irrigation and drainage organisations must be able to formulate and setup adequate management systems and strategies and implement management plans with clear strategic objectives. This requires fully integrated financial planning across the organisation. All this has to be accompanied by a change in attitude of all the members of the organisation from an administrative government orientation to a responsive client orientation.

3.2   MANAGEMENT FUNCTIONS

Management is defined as the attainment of organisational goals and objectives in an effective and efficient manner through planning, organising, leading and controlling the organisational resources (Daft, 1995).   These management functions form part of the overall management process, which the organisation must implement to focus its activities on the delivery of the organisation objectives. (fig. 3.1).

Figure 3.1   The management cycle (Daft, 1995)

Planning is the development of a blueprint for the achievement of goals and objectives. Some authors distinguish various levels of planning: operational, tactical and strategic in an increasing order of generality and time span. It defines where the organisation wants to be in future and how to achieve these objectives. Typically, this process is called "strategic planning" or "corporate planning". Goals and objectives with their respective targets are defined together with the associated strategies and development plans leading to their achievement.

Organising consists of the strategy developed to accomplish the management plan. It involves identifying the tasks to be carried out, by whom, the management of staff and the allocation of resources to accomplish the planning goals and objectives.

Leading is concerned with providing clear directions to managers and workers for attainment of the management plan and is a critical management function. Managers must be trained to communicate, direct and select the most appropriate approach to ensure that everyone in the organisation 'owns the plan'.

Monitoring and control activities help to keep the organisation 'on track' towards achieving the organisational goals and objectives, and correcting any significant deviations as needed.

The integration of these activities in a management cycle is shown in Figure 3.1. The management functions described above form part of a management process that is designed to transform inputs into outputs. Inputs are the resources the managers have at their disposal to carry out the organisation's plan as efficiently and effectively as possible. In irrigated agriculture this concept can be applied at two levels: (a) the agricultural production system where water becomes one of the inputs into the productive process; and (b) the organisation's management system where resources are used in the process of delivering water supply and drainage services. The main types of inputs are human, financial, technological, and in the case of irrigation and drainage also natural resources. Outputs as products and services are defined in the goals and objectives that the organisation is designed to provide.

3.3   IRRIGATION AND DRAINAGE SYSTEM MANAGEMENT

As described earlier, irrigation and drainage systems must be managed to allow the users of the services to achieve their objective of optimum productivity of their agricultural activities. The management of irrigation and drainage entails some distinct characteristics, which have led various authors to propose several definitions in the past.

Lowdermilk (1981) defines irrigation management as "*...The process by which water is manipulated (controlled) and used in the production of food and fibre*". The author also stresses that it is not water resources, dams or reservoirs to capture water; nor codes, laws or institutions to allocate water; nor farmers' organisations; nor soils or cropping systems. It is, however, the way these skills and physical, biological, chemical, and social resources are utilised for improved food and fibre production. This definition is specifically focused on the process of water management itself rather than on the management of resources involved in the delivery of irrigation and drainage services. Lenton (1986) on the other hand defined irrigation management as: "*.....The process in which individuals set objectives for irrigation systems, establish appropriate conditions, and identify, mobilise and use resources so as to attain these objectives, while ensuring that these activities are performed without causing adverse effects.*" This definition emphasises the achievement of the project objectives set out for a particular irrigation system as the key element that drives management of irrigation. However, irrigation system objectives emphasised in this definition are primarily of social and economic nature, e.g. poverty alleviation, food security, etc, rather than of a management nature.

A common feature of these definitions is the sole focus on irrigation activities with no explicit reference to drainage. It is very common, especially in humid tropical regions that the provision of drainage services is as important or even more important than the management of irrigation services and often both activities are the responsibility of the same organisation. However, drainage services are of a more varied nature than irrigation services. They differ from irrigation services in several ways as illustrated in Table 3.1.

Table 3.1 Differences in properties between irrigation and drainage services

Property	Irrigation	Drainage
Nature of transaction	Economic good + service	Service
Desirability of water	Wanted	Excess water unwanted
		Water for water level control wantedd
Types of customers	Single-specific	Multiple- not specific
	(agriculture, although may include other sectors)	(rural, urban, industrial)
Predictability	Predictable (deterministic)	Unpredictable (stochastic)
Water quality	Client wants highest quality standard for use	Client wants lowest quality standard for disposal
		Clients want good water quality for water level control

Irrigation services are concerned chiefly with the delivery of water for agricultural purposes, although irrigation canal networks are increasingly being used for delivery of water for other purposes such as urban and industrial use. In all cases the transaction involves both water, itself an economic good, and the associated delivery service. Unlike irrigation, drainage despite being necessary relates to water as an unwanted good to be disposed of. Moreover, the timing and amount of drainage, especially from surface runoff is beyond control of both the service provider and the service recipient.

The quality of the drainage water may be affected by a variety of pollutants originating from irrigation or other activities. Compliance with environmental regulations often makes it necessary for drainage organisations to monitor and enforce effluent quality standards.

We consider that a broader definition encompassing both activities is warranted to reflect their distinct characteristics. In this context, we propose a more comprehensive definition of irrigation and drainage system management. *"Irrigation and drainage system management is the process by which resources are allocated and used to provide irrigation and drainage services in a sustainable and cost-effective manner".*

This definition integrates the management of irrigation and drainage systems. It focuses on the management process involved in making use of various resources needed to deliver irrigation and drainage services; the emphasis is therefore on the process that transforms resource inputs into outputs in the form of irrigation and drainage services.

The scope and the scale of the irrigation and drainage system management activities differ for each different irrigation and drainage authority. However, the main processes managed within an irrigation and drainage service can be grouped into three main categories according to their focus: water, structures and organisation[6] (Uphoff, 1991). Table 3.2 provides a summary of these main activities in each category. The activities centred on water deal with the tasks of acquiring, allocating and distributing water for supply and disposal of excess water. The activities centred on structures deal with the tools required to perform the water related activities, while those activities centred on organisation deal with the development and management of the infrastructure to provide the irrigation and drainage service.

Table 3.2 Irrigation management activities (after Uphoff, 1991)

A	Activities centred on WATER
	♦ Aqcuisition of water (surface/groundwater)
	♦ Allocation of water (water -use- rights)
	♦ Distribution
	♦ Drainage (collection, disposal)
B	Activities centred on STRUCTURES
	♦ Planning and design
	♦ Construction
	♦ Operation
	♦ Maintenance
	♦ Renewal
C	Activities centered on ORGANISATION
	♦ Decision making
	♦ Resource Mobilisation
	♦ Communication and information
	♦ Conflict resolution

3.4   THE STRATEGIC PLANNING PROCESS

The main aim of the strategic planning exercise is to enable the organisation to meet the needs of its customers, overcome challenges and take advantage of opportunities arising from changes in its environment. Appraising the agency's environment is therefore a key step in the planning process that will enable management to identify the strategic issues to be later addressed by the organisation. The process becomes one of critical review and assessment in which alternatives are evaluated against well-defined criteria. The development of these criteria also forms part of this process. The main output from this analysis is a definition of strategic objectives, strategies and actions, with reference to targets, milestones, indicators of performance and possible resource usage.

Strategic planning is a process of thinking and decision making which involves not only internal relationships but also those between the organisation as a whole and its

---

[6]   The irrigation management activities listed by Uphoff only refer to the main system facilities and do not include on-farm irrigation management activities.

"transactional" environment (Ackoff, 1981). The general process involved in the formulation of a strategic plan consists of the following:

- review of the organisation's purpose;

- inventory of the transactional environment in which the organisation must perform its current position, functions, risks, challenges and constraints;

- identification of realistic, achievable but challenging objectives and goals with time based targets;

Box 3.1 . Advantages of Strategic (Corporate) Planning

- Greater purpose and direction for the whole organisation

- Sharing of common goals by the various component units

- Agency sensitivity to the external environment

- Identification of strategies to react to that environment

- formulation of strategies, programs and activities directed towards achieving the agreed objectives and goals;

- inventory of the resources the organisation has at its disposal including their strengths and weaknesses;

- formulation of the actions plan that results from this analysis.

- Monitor and evaluate progress towards achieving objectives and goals.

Corporate-strategic planning can assist an organisation to respond more flexibly to the development of a service oriented operation because it allows for a wide consultation process to take place and responds to often-conflicting pressures (Box 3.1). Figure 3.2 shows the main activities involved in integrated planning for irrigation and drainage agencies.

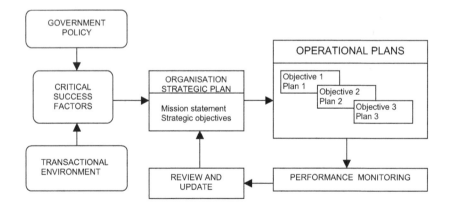

Figure 3.2   Conceptual framework for integrated strategic planning

3.4.1 The Transactional Environment

The assessment of the *environment* surrounding irrigation and drainage is the first and most essential task in the development of the strategic plan. It explores the context in which the plan is set, the existing challenges, risks and constraints. The environmental analysis anticipates specific future directions and guides, the incorporation of new policies and programs and the discontinuation of old ones.

There are several parts in this exploratory stage. The tasks involve a broad review of reasons why the agency exists. For example what is the business of the organisation and what is the policy and legal framework within which the organisation must function? What is the history of the agency and what are the values and ideals that should be preserved by the organisation? Finally, what are the new directions and perspectives that the agency should adopt to respond to its customer's needs and to government policy? Answering these questions will rely on availability of information related to the past performance of the organisation and the constraints and boundaries within which the organisation will have to function. These are system-specific issues that will then provide the focus for the formulation of the specific goals and objectives of the plan. The output will be a broad appreciation of the current and future issues confronting the irrigation and drainage organisation (Box 3.2)

Box 3.2 Strategic Planning Critical Questions
◆ Where are we now?
◆ Where do we want to be?
◆ How and when do we get there?

The most notable outcome of this task is the identification of *the critical success factors (CSF)*. These are factors that address issues critical in determining the outcome of the irrigation and drainage activity. The identification of these factors normally follows intense consultation between the agencies and stakeholders either involved or connected to the management of irrigation and drainage systems.

Typical examples of these factors are: the management of the land and water resources in a sustainable manner, provision of a timely and reliable delivery or water to irrigators, improvement of poor performance of irrigation assets, etc.

3.4.2 Organisation's Purpose

A strategic plan contains a hierarchy of statements of purpose, which are usually stated as mission statement, goals and objectives, all relating to a vision about the future of the organisation. The first task of senior management in the organisation is to formulate the ideal end-state of the organisation that it wants to create for the future. This is called the *mission statement*. It is a short succinct statement that set outs the essential purpose of the organisation (Box 3.3). In formulating its mission, the organisation must answer four main questions (Goodstein, *et al.* 1993):

- ◆ What function(s) does the organisation perform?
- ◆ For whom does the organisation perform this function?
- ◆ How does the organisation go about filling this function?
- ◆ Why does this organisation exist?

The purpose of the mission statement is to provide a clear and common understanding of the organisation's role both within and outside the organisation. It describes a very general purpose of the organisation, which generally integrates the variety of roles that the organisation plays. The mission statement provides cohesiveness of purpose and a focus to plan the agency's operations in a comprehensive and integrated way. It must describe what type of service the organisation wants to provide and to whom it wants to provide it.

Box 3.3 Mission statement of Goulburn-Murray Water, Victoria, Australia.

Goulburn-Murray Water will deliver price efficient, sustainable water services to Northern Victoria.

3.4.3   Strategic Objectives

The *goals* of the strategic plan must reflect the priorities identified in the critical success factors. They must be stated as explicitly as possible as they will become the strategic guide to the management process. Specific *objectives* and *targets* must be attached to each goal to provide a quantification of what is to be achieved. The time dimension is an essential element that must be added to the quantity or quality (norm) of a performance indicator that will be used to indicate progress "along-the-way" towards accomplishing the established objectives.

3.4.4   Operational Plans

The main product of the strategic-corporate planning process is a management plan that provides the blueprint towards the achievement of the plan main goals and a rational basis for the formulation of forward and annual budget estimates. This plan usually encompasses a number of elements, which address each individual goal.

In addition to these specific development plans to address individual goals, additional plans are needed to support their implementation including a financial plan, human resource development plan and information systems plan. Collectively they all form the basis for annual budgets and financial planning.

The outcome of the strategic planning process therefore will be an operational plan, which consists of a set of plans:

♦ a *water management plan* dealing with the basic policy on acquisition of water, water rights and priorities of water rights, water allocation procedures, water distribution, drainage and water quality standards for irrigation and drainage;

♦ an *asset management plan* dealing with the physical infrastructure (flow control systems, structures, canals, equipment and plant) required to provide the irrigation and drainage services;

♦ an *organisational management plan* dealing with the division of tasks and responsibilities, the financial management process, the management information systems, and the development of human resources.

New activities and services identified in the strategic plan will require special staff skills that may have to be developed through an appropriate training program. For instance, if an improved level of service requires increased flow monitoring capacity, operatives involved in the field operation would have to be trained accordingly. The human resources development programs must always form part of the strategic planning and must be designed to support the plan's strategies that are implemented towards the achievement of the plan's goals.

For each element of the strategic and operational plans, specific indicators will provide the most useful benchmark against which changes or progress in strategic directions are monitored, evaluated and reviewed.

3.4.5 Performance Monitoring

Monitoring of progress in the plan and adjustment of strategies must be an on-going management activity. The program structure provides the first focus for monitoring. Usually targets, indicators and milestones are designated within a program structure for monitoring and reporting progress.

Monitoring should be regular and planned, so that those responsible for delivery can prepare and provide appropriate information for adaptive responses. Furthermore, the process of delineating critical success factors, objectives and goals is of a dynamic nature. Changing agricultural, economic and environmental conditions surrounding the irrigation and drainage activity requires their continuous review. A monitoring and evaluation process that includes an effective feedback mechanism is essential to this review process to enable planners and managers to adjust goals and objectives to mach the changing importance of issues over time. A more detailed discussion on performance monitoring of irrigation and drainage services is provided in Chapter 7.

All these plans require financial resources. Financial planning is therefore an integrating and often decisive activity. In the next section some general aspects of financial planning in irrigation and drainage authorities will be highlighted.

3.5 FINANCIAL PLANNING

Traditionally, irrigation and drainage agencies have operated in an environment in which the level of expenditure on infrastructure bears no relation with the water delivery service that they are expected to provide. Very seldom have agencies clearly defined their level of service and the cost associated with it.

Therefore, the amount of funds available for the provision of service is often not commensurate with performing an adequate program. Underlying this approach is the notion that the level of service and consequently the level of expenditure are determined by inputs such as availability of money or other resources. Such *input* driven budgetary process focuses on the allocation of inputs and resources in accordance with prior commitments with little or no information on, or attention directed to performance. The operation and management of irrigation and drainage systems with a service orientation entails an *output* driven approach. To achieve this, management needs to know what is an acceptable standard of service and what are the costs and resource implications of providing that standard of service. In other words, the management of resources must be consistent with the definitions of management goals and objectives (Lindley, 1990).

A key element of financial planning is the ability to ascertain the resources that are needed by the courses of action and plans that have been selected including money, infrastructure and human resources. The process involves the determination of the resources that are necessary to implement the plan and how much will be available to the organisation considering its predicted revenues. The amount of resources that must be generated or acquired can then be determined.

Figure 3.3 provides a framework for the financial management process (van Hofwegen, 1997). The cost of service provision is directly related to the level of service provided. The higher the level of service, the more management inputs and infrastructure are needed, hence a higher cost is incurred. In situations where customers pay the full cost of service provision, the level of service must be balanced against their willingness to pay through a process of consultation. The result of this consultation is then reflected in a service agreement between the organisation and its customers.

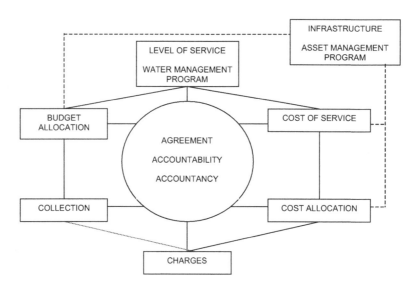

Figure 3.3 Conceptual framework for financial management (after : van Hofwegen, 1997)

Cost allocation procedures must be introduced in agencies providing more than one service, with the cost of each service clearly identified. The level of charges or service fees can then be based on the overall cost of service while cross subsidies may occur internally across services.

Decisions on tariffs are usually linked to government regulation and policy due to the monopolistic nature of these services. Often tariffs are fixed or government sets ceilings. If these are insufficient to cover the actual cost of service provision, subsidies are required for a sustainable level of service.

If the organisation has financial autonomy, service charges can be collected directly by the agency. Otherwise, charges are collected and reverted into general government revenues, in which case clear guidelines must exist to guarantee that funds are reallocated to the managing organisation to ensure the adequate provision of the agreed level of service.

The collection can be linked to the delivery of services directly or indirectly through a system of land or other taxes.

Budget allocation to various activities is the final stage of the financial planning process. The main outcome of strategic planning is a set of integrated plans, which may include plans for development, asset management, research and investigations, human resources development etc. Clearly the mechanism to allocate and reallocate resources can only be effective if it is based on a program that reflects the purposes and priorities of the corporate plan. Because strategic plans are of a dynamic nature, financial planning usually results in "rolling plans" which may forecast activities for a period ahead, generally 3 to 5 years. This process involves annual updating based on a review of the last year's performance and analysis of future needs.

Effective financial planning therefore requires the ability to forecast the financial position of the organisation for the selected planning period. Financial modelling can be a very effective tool to assist with this process. Modern information technology systems can be very effective in enabling management to integrate all the resources across different areas of the organisation.

In the preparation of the annual budgets and forward estimates it is essential that information from the planning process is available, so that the resource allocation accurately reflects the organisation's priorities.

The budget process itself has several distinct phases, often referred to as the Budget Cycle. This cycle includes:

- formulation of budget estimates;
- the approval process;
- implementation and monitoring of approved programs;
- review of completed programs.

For an organisation to gain maximum advantage from its program budgeting, it should have a means by which performance measures can be established and monitored on an ongoing basis. This entails a review capacity whereby information on progress and achievement towards objectives can be gathered and assessed against predetermined criteria and targets.

Chapter 4

The Irrigation and Drainage Service Concept

One should not talk about a "rule" unless most people whose strategies are affected by it know of its existence and expect others to monitor behaviour and to sanction non-conformance

E. Ostrom, 1993.

The rationale to consider the activities of an irrigation and drainage organisation as those typical of a business organisation was elaborated in Chapter 3. This chapter deals with the nature of the transactions carried out by irrigation and drainage organisations and is aimed to address the nature of the business and the elements and activities that form part of conducting the service business. The type of service provided by irrigation and drainage organisations is determined by a number of physical and socio-economic factors. There are four attributes of the level of service that determine its quality: adequacy, reliability, equity and flexibility. The level of service for irrigation and drainage systems consists of a number of specifications and conditions, which must be formalised through proper service agreements between the service provider and users within a framework of proper accountability.

4.1 IRRIGATION AND DRAINAGE AS SERVICES

Irrigation and drainage agencies in many countries are entrusted with the responsibility of operating the irrigation system to supply water to farmers and to remove excess water through the drainage system. In general, very few agencies maintain responsibility for operating the irrigation network down to the farm level although the interface where the agency's responsibility ceases and the farmer's responsibility begins varies in different countries and within countries. Regardless of the level of agency responsibility in managing the system or the type of organisation, the main transaction involving the agency and customers (users) is the supply of water for irrigation, protection against floods, the removal of excess water or a combination of these. These constitute the main *business* of the irrigation and drainage organisation. The *products* involved in the business transactions are water supply, flood protection and water removal. To carry out its business, the organisation relies on hydraulic infrastructure for the acquisition, storage and conveyance and control of flows and (ground-) water levels for water supply and removal and protection against floods. This infrastructure consists of assets such as dams, canals, control structures, pumping stations, dikes, etc., that enable the water control functions to be carried out.

Before looking into the concept of irrigation and drainage services, first the concept of service must be defined. Kotler (1994) defines service as *"…An activity or benefit that one party can offer to another that is essentially intangible and does not result in the ownership of anything. Its production may or may not be tied to a physical product".*

This definition highlights the concept of intangibility of the activity involved and the fact that a physical economic good may be part of the transaction. In the case of irrigation services the activity involved is water supply whereas in the case of drainage services the main activity is water removal. These services are connected to a physical economic good -water- If follows that distinction must be made between the nature of the good and that of the service. Kotler further identifies several categories of good and service combinations in products ranging from pure tangible goods such as jewellery, furniture, etc, to pure intangible services such as medical services, psychotherapy, rubbish collection, etc. In this context, in the irrigation water supply service both a service and an economic good are involved whereas drainage only involves an intangible service.

The actual composition of the product of irrigation and drainage authorities recognises two elements:

- ♦ the *service* consisting primarily of acquiring, storing and conveying water from the source to the customer (the irrigator) and/or the collection and disposal of drainage water;
- ♦ a tangible *economic good* represented by the water delivered to the user.

While the notion of the delivery service and its associated cost is often easy to grasp, the notion of water being an economic good is less apparent. There are a number of cultural, religious and social aspects that policy makers must often consider in addition to economic efficiency in the formulation of water pricing policies. The international community during the International Conference on Water and the Environment in Dublin (ICWE, 1992) asserted that *"….it is vital to recognise first the basic right of all human beings to have access to clean water and sanitation at an affordable price".* The implication of this statement is that water is essential to life and access to it is a basic human right that requires water pricing and allocation policies that are not driven solely by market considerations.

Irrigation water is considered an economic good with an economic value and it is an important input into agricultural production. Its value however varies in time and place. The value is highest for the farmer that needs to irrigate to prevent yield reduction. However, for the farmer next door who just finished irrigation or after a rainstorm, irrigation water may be of no value and may be either rejected or directly bypassed to a drain. In drainage practices water can be considered as having no value even though service transactions are involved. Through drainage, water of no value is removed from land of the customer who considers it harmful to his/her economic activities and/or quality of life. If drainage water is reused for irrigation purpose it may acquire again an economic value.

Translating the economic value of water and the cost of service provision into a water pricing and allocation policy is the concern of policy makers. Such pricing policies are not only designed to cover the cost of service but to meet the basic social and human needs. Once government has adopted a policy, it then becomes a part of the environment in which the irrigation and drainage authority must formulate its strategic plan and define the characteristics of the irrigation and drainage services. The pricing of water as an economic good is of critical importance to the organisation providing the water supply service since it would often become a significant component of the total service price. If the drainage organisation is only responsible for the provision of a service consisting of the removal of excess water, the service component is clearly more dominant because there is no delivery of any tangible economic good in the provision of drainage service. This concept also applies to the provision of flood protection to urban and agricultural areas.

4.2   DEFINITION AND QUALITIES OF LEVEL OF SERVICE

The key element to be addressed in a strategic plan is the management and operation of the irrigation and the drainage system for the provision of a specified level of service. This level of service must be specified in the form of a set of quantitative operational standards that can be used by the organisation to guide the supply of irrigation water or the removal of drainage water. Meeting these standards at minimum cost then becomes the main objective of the organisation for which the necessary operational objectives, rules and procedures must be formulated (Lee *et al*, 1998). These specifications then become the norm against which the operational performance of the organisation can be measured.

The level of service must emerge from an extensive process of consultation and must be agreed upon between the irrigation and drainage organisation as the service provider and the farmers and other clients. The Government[7] and the river basin authority also participate in this consultation process as supervisory and regulatory bodies.

The concept of *level of service* is defined accordingly as (Hofwegen and Malano, 1997) *"….A set of operational standards set by the irrigation and drainage organisation in consultation with irrigators and the government and other affected parties to manage an irrigation and drainage system"* This definition implies that the formulation of the level of service criteria must contain several qualities. The level of service specifications must be consistent with the goals and objectives identified in the strategic plan of the organisation. This process as discussed in Chapter 3 must involve an extensive consultation between the organisation providing the service and the users. To be consistent with the strategic goals of the organisation, there are some specific attributes and qualities that must be attached to the formulation, implementation and review of the level of irrigation or drainage service provided (Box 4.1). As part of the formulation process, performance indicators, standards and targets are identified to enable assessment of compliance and level of performance in the delivery of service.

[7.]   The government will not always be directly involved in the level of service negotiations. In such case, the role of the government is to provide a framework within which these negotiations between service provider and customers can take place and to ensure that society interests are taken into consideration.

The opportunity for formulating service specifications varies according to the time that the development and implementation of the strategic plan occurs in relation to the development of the scheme. Two situations may occur: (a) the level of service is specified and agreed upon before the system planning commences[8], and (b) the level of service is specified for an existing system. In the first case, the designer must refer to these specifications to determine the appropriate type of hydraulic control that together with the appropriate management are required to meet the level of service. Otherwise, the level of service specifications must be developed in accordance with the existing hardware infrastructure if the system is already in operation.

Being a service activity, irrigation and drainage must reflect the changing nature of the main economic activities that it supports. In the case of irrigated agriculture, this is agricultural production whereas for drainage services it is a range of activities that are protected from waterlogging and flooding including agricultural, urban and industrial areas. An example of such changes is that farmers in many rice based irrigation schemes have moved towards crop diversification requiring a more flexible irrigation water supply. Moreover, upland crops demand better drainage protection for which an increase in drainage capacity is needed. In such situations, the set of service specifications that was appropriate for rice production is not likely to suit the requirements of farmers growing upland crops. The level of service must evolve to respond to these changes. Moreover, as drainage service is often provided for a combination of uses like agricultural and urban areas the drainage service specifications may have to satisfy the needs of all these clients. This also implies that the clients of drainage services are most likely not the same as the clients of irrigation services.

The cost associated with the provision of a given level of service is another key aspect to be discussed during the consultation with users. A firm link must be established between the provision of a certain level of service and the concurrent service cost. In this process, all the cost incurred in the provision of the service must be explicitly identified so that the relative importance of each cost component becomes transparent. This is even more important for organisations providing more than one service to different beneficiary groups. The actual cost of service delivery can then be properly ascribed to each service.

> **Box.4.1 Important Qualities of Level of Service**
>
> ♦ it must emerge from an extensive consultation process
>
> ♦ it should become a series of norms (targets) against which operational performance is measured
>
> ♦ it must be revised on an on-going basis to respond to changes to irrigated agriculture
>
> ♦ it requires careful consideration of the cost associated with the provision of specific levels of service.

It is important at this stage to establish the difference between the cost of service provision and the level of cost recovery. Determining the cost of service is typically the role of the managing organisation, while decisions about the level of cost recovery from users is the responsibility of Government. Governments often make explicit policy decisions to

---

8. Determining a level of service in a consultative process before establishing a scheme may sometimes be difficult if the future users are not familiar with irrigated agriculture. In the process of service development a gradual development of service levels should therefore be considered

subsidise the irrigation and drainage activities on the basis of supporting the agricultural sector either for strategic or political reasons. However, it is important that the costs associated with the provision of the irrigation or drainage service are carefully identified so that the level of needed subsidy can be properly ascertained.

A comprehensive analysis including all cost items provides the basis for identifying all the costs attributable to the provision of services where all assets are formally valued. This approach enables management to include the rate of depreciation or asset consumption in the calculation of the overall cost of service, and makes explicit the relation between level of service provision and actual cost. A more detailed discussion of cost of service provision will be provided in Chapter 6.

The operation costs consist primarily of personnel, infrastructure, plant and equipment required for the operation of the system. The relative importance of the operation cost depends on the intensity of field operation activities involved in the regulation of water control and service delivery to users. In general, where the interface between agency and users occurs at a high level in the system, e.g. main canal servicing secondary canals operated by users, the staffing requirement is less than in systems in which the interface occurs at a lower level in the system. The degree of water control automation also has a significant effect on the operation cost. More highly automated systems require fewer staff to carry out the operation tasks. However, these types of systems are more capital intensive. The higher capital cost can easily offset the reduction in staffing cost. A more comprehensive discussion on the relation between cost of service provision and infrastructure and management inputs will be provided in Chapter 5.

4.3   LEVEL OF SERVICE DETERMINING FACTORS

Irrigation and drainage organisations operate under a physical, legal and socio-economic setting that defines the environment in which the level of service specifications must be developed. The difference in nature between irrigation and drainage services requires the separation of the discussion on their determining factors and service specifications.

4.3.1   Determining Factors for Irrigation Services

For irrigation agencies, the formulation of level of service specifications is a process of "customisation" of the irrigation supply parameters duration, flow rate and frequency. There are several factors that form the boundary conditions of the customisation process and the final service specifications. These include:

- crops, soils and climate;
- systems hydraulic control infrastructure;
- water resources;
- sediment load;
- irrigation practices and farmers skills;
- water rights;
- irrigation policy;

- ♦ competing water uses;
- ♦ environmental regulations.

The same factors affect the development of the level of service in new and existing irrigation schemes although relevance and interaction between the various factors differs. This analysis is based on the premise that new irrigation schemes are developed to meet clearly defined level of service specifications.

Crops-Soils-Climate

The combination of crops, soils and climate determines the basic water requirement and the preferred form of on-farm water application and consequently irrigation supply. The climatic characteristics of a region determine the degree to which the crop water requirement must be met by irrigation. In arid areas, irrigation plays a more critical role not only in the water supply but also in driving important farming decisions regarding planting dates and other cultural operations.

The cropping pattern affects the definition of level of service in relation to water demand, frequency, and economic value of different crops. Crop water demand is a function of climate and varies with different growth stages. These together with the soil water holding capacity and root depth will determine the desirable frequency of water application. Crop value and sensitivity to water stress play an important role in the degree of reliability of supply that irrigators may desire from the irrigation system. Rice based cultivation does not require elaborate level of service specifications and is more tolerant to less flexible delivery schedules. On the other hand, vegetables require more frequent irrigation and are very sensitive to water stress requiring more security of supply and a more flexible delivery schedule that must be met by a higher level of service.

In defining the level of service specifications, it is not only the actual crop water demand but also the farmer's preferred form of irrigation delivery that must be taken into consideration. Irrigation is only one of many tasks that the irrigator must carry out in order to grow a successful crop. In addition to soil manipulation, fertiliser and chemical applications, irrigation often must be "fitted-in" with these tasks which may have a higher priority with farmers.

Hydraulic Infrastructure

Flow control systems are a critical component of the hydraulic infrastructure that determines the ability to provide a specified level of service. The role of the flow control system is to control the discharge and water level at flow division points to meet the level of service specifications.

Canal discharge and water level can be regulated by several means including (a) water level control, (b) discharge control, and (c) volume control. Most systems use a combination of water level control and discharge control. Discharges through offtakes are often indirectly controlled through the water level in the parent channel. Hence, the variation in discharge at the off-take is the result of variations in the upstream water levels and the sensitivity of the combination off-take and water level regulator to those variations. Fluctuating water levels in canals require concurrent adjustments of the gate opening to maintain the discharge specified by the level of service. Inappropriate selection of canal controls can yield hydraulic systems that are unstable and very sensitive to discharge variations. Such systems are difficult and impractical to control in order to meet a specified level of service. If the provision of service requires stable discharge delivery, it is highly desirable to have a system that can maintain a stable water level. There is a wide range of hydraulic control systems that can be used to achieve specific water control objectives in an irrigation system. These will be described and analysed in more detail in Chapter 5. Here only the general features of water level regulation in upstream, downstream and volume control systems are discussed.

Upstream control is by far the most commonly used type of infrastructure around the world. With upstream control, the water level is monitored upstream of the structure that controls the water level. Prior to a change in discharge a change is required in the setting of all the control structures upstream of the location concerned. Upstream control systems are not well suited to respond to continuous changes in demand. They are rather better suited for supply oriented operations with limited flexibility such as imposed or scheduled deliveries. Downstream control responds to water level changes downstream of the regulator. Downstream control is well suited to respond to changes in demand downstream of the point concerned. With proper instrumentation and automation, they can be designed to provide an instantaneous response to changing demand conditions. Volume control systems involve the simultaneous operation of all control structures in the system to maintain a nearly constant volume of water in each canal pool. With this type of operation, the water surface in each canal pool rotates around a point located approximately midway between the control structures. The implementation of volume control requires the automation of the control structures coupled with centralised monitoring and control.

In addition to the type of hydraulic infrastructure, the ability to provide a certain level of service depends on the available management and operation capacity. In theory, any level of service can be delivered by both manual and automatic control systems provided that adequate operation and management inputs are available. This assumes that substitution can occur between hardware and software to provide a given level of service. In practice, however, there are limits to this trade off. Experience shows that the level of management and the operation skills are often the main constraints to deliver a flexible level of service with manually controlled systems. Downstream automatically controlled systems allow a greater level of flexibility and require fewer staff albeit with a higher level of expertise in operational and maintenance planning.

Water Resources

The reliability and availability of the proper quantity and quality of water is an important factor influencing the formulation of level of service specifications. Regulated systems provide an increased level of reliability and flexibility by using reservoir systems that are designed to supply water during dry periods when natural stream flows are lower than normal.

The potential of irrigation depends largely on the constraints to availability of water resources. This is why the potential of existing systems or design of new water supply systems is usually based on high flow security. For regulated systems, the total shortfall in reservoir inflows during the critical period determines the minimum volume of storage required or the maximum level of demand, which can be satisfied through that same period. Such a storage-yield analysis is undertaken taking into account an agreed set of performance standards or drought security criteria for the system.

The inherent large variability and unpredictability of stream flows means that there will always be a certain risk of a drought period occurring during which the water supply system will not be able to fully meet demands. The objective of the water supply system analysis is to determine the sustainable level of supply to demands in the system for which an agreed set of performance standards can be met. This is normally termed the safe yield or firm yield (Linsley & Franzini, 1979). To determine the firm yield a study must be carried out of the relationship between storage size, sustainable level of water demands (or yield) and reliability of supply. The storage behaviour is simulated for different storage capacities and water demand of different and often competing types of water supply systems placed on the same stream. The reliability of reservoir supply is defined as the probability that a reservoir will deliver the expected demand throughout its lifetime without incurring a deficiency.

The reliability standards for supply need to be specified with regard to the adopted level of risk of not being able to supply water according to a specified performance standard in any one year. This supply reliability will then become part of the level of service specifications related to the expected allocation and availability to users. It will also form the basis to delineate the restriction policy that will be applied in case of water shortages. Reservoir rule curves are often used to determine "trigger levels" at which restrictions are applied. The restriction policy may differentiate between different user groups so that the severity of restrictions on supply to different users reflects their ability to cope with shortfalls in supply.

Many existing water supply systems are capable of providing a very high level of reliability of supply in the early stages of development because their design allowed for an expected future growth in water demand for irrigation, urban and industrial use. This high level of reliability which users experience in the initial stages of development may give rise to expectations of a continuing high level of reliability which in most cases is not possible to sustain. It is therefore important that the drought security criteria that were adopted are clearly spelled out at the beginning of the development process so that the various users groups are made aware of their implications regarding possible restrictions on supply in future. The fact that the actual reliability of supply changes over time underlines the need for an on-going review of the level of service specifications. This is illustrated by the fast rate of siltation of many reservoirs leading to reduced reservoir capacity and supply reliability.

In regulated systems, the irrigation agency can advise irrigators on the availability of water before the irrigation season begins to enable them better planning of their agricultural activities. Whenever this occurs, it is critical that such process is clearly spelled out in the level of service specifications as to the timing of the announcement and the amount of water allocation available for the season.

Unregulated water supply systems are more severely affected by variations in the natural river runoff. Level of service specifications must take into consideration the wide discharge fluctuations that are likely to occur by establishing a clear set of criteria relating to water supply and water availability. In general, the ability of unregulated systems to provide a flexible irrigation supply is limited and this will be reflected in the level of service specifications.

Use of surface water is usually the most important resource for irrigation and drainage authorities. Use of groundwater is usually organised and managed by farmers themselves, although in some arid regions groundwater schemes are under the management of the irrigation authority. In such schemes the water availability can be compared with that of a surface reservoir. The abstraction out of the aquifers should not exceed the safe (sustainable) yield if mining is to be avoided. Reuse of drainage water or wastewater can be treated like additional surface water although its use must meet irrigation and safety water quality criteria concerning salinity, chemical and biological contamination levels.

Sediment Load

In many run-of-the-river schemes sediment load is a major factor in the management of the irrigation system. Sedimentation in irrigation canals remains a major concern despite the fact that many techniques have been developed to remove sediments at the system intake. In many countries, especially in the humid tropics, sediment management procedures prescribe the maintenance of a minimum flow to keep sediment in suspension to minimise sedimentation. In periods of low flows, this requires augmentation of discharge through the introduction of rotations between lower order canals. Such rotations limit the level of flexibility of water delivery. The alternative for such rotations would be increased maintenance cost of desilting the canals.

The choice of flow control systems in canals with high sediment loads can be important in determining the level of maintenance cost. Systems where minimum water levels must be maintained to obtain proper supply through the offtakes are usually prone to higher sedimentation rates than systems where distribution does not rely on controlled water levels (e.g. proportional systems). The highest sediment trapping capacity will be found in downstream control systems. This is often sufficient reason to avoid this option where water has high sediment loads. This is another case where the level of flexibility in water delivery must be balanced against the additional cost of desilting the irrigation canals.

Water Rights

The right of access to water for agencies and individuals is in general established by legislation. There are a number of water right systems in the world; however, they all fall into two main doctrines: (a) the riparian doctrine, and (b) the appropriation doctrine.

The riparian doctrine is based on the common law of England. The main tenet of this doctrine is that landowners contiguous to a stream are entitled to access the stream flow undiminished in quantity and unpolluted in quality. Ownership of land overlying an aquifer is sufficient to establish right to ground water. A strict application of this rule would not permit consumptive use of water from a stream or well for any purpose (including irrigation). Because of this, nations where the riparian doctrine is applied have introduced modification to their water law to attenuate the right to access.

The appropriation doctrine asserts that water is a public good, but a right to access may be obtained by individuals or agencies provided they comply with certain requirements and principles. Beneficial use is the main criterion to limit the right to access and maintain the water rights. Although beneficial use implies that the water entitlement must be used, this does not necessarily imply that it is used in a productive and efficient manner. Rights may be lost by failure to meet this beneficial use criterion. In some cases, priority rights of appropriators must be satisfied before subsequent appropriators can access water. Other variants of the doctrine establish priorities of use on the basis of the type of use, e.g. municipal, industrial, agriculture, etc; or on the basis of crop types, e.g. perennial fruit trees are assigned higher priority because of their long establishment period and capital investment required. The elements of an appropriation right include quantity of flow, time of use, point of diversion, nature of use, place of use, and the priority of right. In general, none of these elements may be changed without prior approval by the competent authority. Failure to do so may infringe on prior rights. These two doctrines provide the general framework for the formulation of the specific water laws or acts. These contain further specifications governing the way in which water is shared between individuals and the conditions that apply to those rights. Murray-Rust and Snellen (1993) identified six principles to establish the right to water allocation:

- share per unit area;
- share per person or household;
- fixed discharge per unit area;
- fixed volume;
- instantaneous demand;
- informal or undefined rights.

This list contains the basic principles for water allocation. There are however a large number of possible combinations that are applied around the world. These rights may be restricted by the application of certain conditions such as the suspension of rights, priority of access and other special measures to address situations of water shortage, cropping policies or share of resource with other uses. Recently, the creation of water markets in some countries such as Chile, Australia and the USA has provided another vehicle for the allocation or reallocation of water entitlements.

The principle for water allocation has significant implications in the formulation of the level of service. Water rights that are attached to area or household are more likely to impose restrictions on the flexibility and adequacy of the water supply by limiting the user's freedom to select the best opportunity to use the water. Volumetric allocation gives the user and agency the opportunity to develop level of service specifications that enable users to exercise their choice in deciding when to use their allocation subject to certain constraints imposed by the hydraulic infrastructure.

A significant improvement in the level of service can be obtained where the water rights legislation allows the transferability of water entitlements. Flexibility and adequacy can be improved as a result of allowing users to transfer their water entitlement either temporarily or permanently. The improved economic efficiency of water is an additional benefit as the actual economic value of water is better reflected in these open market transactions. However, the transfers of water rights must be well planned and administered to enable water managers to plan effectively not only at the sub-sector level (eg. irrigation) but also across sub-sectors at the river basin level. This includes the ability to guarantee access to water to meet basic human needs to all members of society and to discourage or prevent speculative transactions.

Farmer Irrigation Practices

The ultimate purpose of irrigation service is to provide users with water for agricultural production. The viability of the agricultural enterprise depends largely on the reliability of the service provided and farmers' ability to pay for it. With a higher level of service, farmers reduce their risks and are better able to invest in farming technology and higher value crops. Hence, establishing and delivering a clearly stated and agreed level of service becomes an important element in the process of economic development.

The concept of irrigation scheduling is to apply water to the crop in the correct amount and at the proper time to maximise crop production and/or profit, while maintaining a reasonably high irrigation efficiency. Scheduling irrigation at the farm level will thus depend on the type of crops, soil and climate as discussed earlier, the irrigation technology and the farmer's skills to apply it. To plan for an efficient use of water, irrigators require information on the actual irrigation needs and the time and amount of water available to them in terms of discharge, duration and frequency. Ideally, irrigators should have direct access to water so they can react adequately to changes in the soil moisture status, optimise their irrigation schedules and synchronise them with other on- and off-farm activities. Consequently, farmers would prefer a maximum flexibility in water delivery. However, in most irrigation schemes this flexibility is difficult to provide, and the amount of water available or the timing of supply is often constrained by the flow control infrastructure.

In developing the water delivery schedule, the irrigation organisation must strive to satisfy the needs of the farmers. Also, the delivery schedule must be compatible with the capabilities and constraints of the flow control system. The success or failure of the irrigation system is largely determined by the supply of water to the individual farmers at the farm gate. In smallholder schemes, where the number of farmers is large, this function of delivery is often performed through Water Users Associations (WUAs) at tertiary unit level. Hence, water delivery requires smooth communication and clear sets of rules and regulations for acquisition, conveyance, delivery and distribution of water. Moreover, an accountability mechanism needs to be in place to verify that these rules are followed. However, these rules are made in different forums. The irrigation agency develops a set of rules related to water allocation and delivery to the users or user groups. The users develop their own set of rules for distribution of water among themselves. These reflect the social and power arrangements among and within farming communities, their water rights and their capability to adjust the distribution of water to their socio-cultural situation. These agreed rules will ultimately constitute the agreed level of service.

Irrigation Policy

Traditionally, the development of large irrigation schemes has been the responsibility of government agencies as part of a broader government development plan. Within this framework of development objectives, irrigation agencies are assigned the task of development the physical and management infrastructure to supply water to users. This interface takes place at some level in the system delineating the responsibility of users and agency. The arrangement of the infrastructure used in most large-scale irrigation schemes involves three levels of operation, namely:

- headworks and main system whose primary functions are to acquire and convey water for supply to the tertiary systems or individual farmers;
- the tertiary systems that distribute water among farmers;
- the farm system where water is applied onto the cropland. This is the point where the actual connection between the hydraulic infrastructure and the production system takes place.

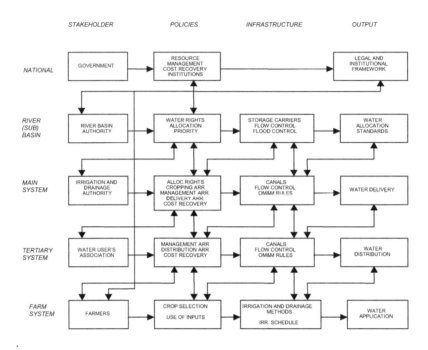

Figure 4.1 Framework for policy choices (Hofwegen and Malano, 1997)

The typical management arrangements that are usually overlaid on the hydraulic infrastructure arrangement consist of the irrigation and drainage authority, a group of formally or informally organised users, and the individual farmers. The irrigation and drainage authority supplies water to a group of users who distribute it among themselves according to certain formal or informal rules.

In this context, we can identify five levels of planning and decision making: (a) the Government, (b) the river basin authority, (c) the irrigation authority; (d) the water users association; and (e) the individual user. At each level decisions are made that impinge on the subsequent level of decisions below in a cascade-like manner as depicted in Figure 4.1. In many countries water user associations and river basin authorities have not been established. Water users associations are absent where the irrigation authority supplies water directly to individual farmers although farmers may be organised in a similar fashion. In some countries the absence of a river basin authority may be due to the fact that either there are no conflicting uses or the irrigation authority has also the responsibilities to manage water at the basin level.

The sequential decision making process involves Government dealing with policy choices involving broad developmental goals that are reflected in its water policy. These policies are translated into a legal and institutional framework for water management including criteria for multiple and often competing uses. In this way, Government provides the enabling environment for the different levels of planning to operate effectively.

If a river basin authority is established, its primary role is to take care of the allocation of water in the basin for different purposes and control or mitigation of pollution of water resources. To this extent, it will either further develop or implement government legislation on sectoral water rights, procedures and priorities for allocations and water quality standards for abstractions and disposal. Policy choices made at the level of the irrigation authority involve among others the formulation of irrigation water allocation rights, water delivery arrangements, cost recovery arrangements and system management. The group of water users can make decisions concerning their internal water distribution arrangements, revenue collection and cost recovery. Individual users can decide on crop selection, use of agricultural inputs, farm irrigation methods and irrigation scheduling. While decision-making involves a two-way interaction, there are often conditioning factors from higher level acting on lower level decision making.

The process of formulation of level of service specifications involving the irrigation and drainage authority and the users can therefore be conditioned by legislation and river basin authority regulations. For instance, government legislation may dictate priority for various competing uses, e.g. domestic, industrial and agricultural. Under these circumstances, the level of service specifications will have to incorporate these constraints which may limit the flexibility of supply. A second set of constraints arises in the translation of the policy decisions into infrastructure and operational rules and procedures to manage the infrastructure. For instance, government water policy may state that water charges will be issued on a volumetric basis. The hydraulic infrastructure should then enable measurement of volume or measurement of flows and duration of delivery and management should record, process and charge for consumption accordingly.

Environmental Regulations

Irrigation and the resulting drainage have an impact on the environment in a number of ways. Irrigation water quality degrades after it is put to use and the consequent drainage flows degrade the quality of receiving groundwater and surface water resources. Surface runoff from irrigation usually shows an increased content of organic debris, sediment load, nutrients and pesticides whereas subsurface drainage effluent usually exhibits an increased concentration of salts and other specific ions. Irrigation can also have a degrading effect on the soils e.g. salinisation if drainage is adequate drainage is not available. Other negative effects of irrigation might be the development of water borne vector diseases like malaria, schistosomiasis, etc.

On the other hand environmental factors such as the chemical and biological quality of the incoming irrigation water can affect irrigated crops, the irrigators and the consumers of agricultural produce

A greater awareness of the environmental consequences of human activities including irrigation is reflected in the various regulations that governments put in place to reduce these impacts. The most common environmental aspects with which the irrigation and drainage activity may have to comply include water quality criteria for irrigation supply drainage effluent standards and water allocation to the environment (environmental flows).

Water quality criteria for irrigation and animal consumption are well documented in the literature (Ayers and Wescot, 1976, FAO, 1992). Salt content and the relative sodicity of irrigation water have negative effects on crop yields and soil properties. Problems with excessive salt content often arise when surface water and ground water are conjunctively used for irrigation supply. Shandying is often used to control the salt content in the irrigation supply by mixing salt water with fresh water. Other elements including boron and chloride while being necessary for adequate plant growth in trace quantities, their concentration in the soil must range within very narrow limits to avoid becoming toxic. The level of service specifications may include references to the upper limits of salinity and toxic elements that are permitted in the water supply. Especially in reuse schemes the specifications may include Biological and Chemical Oxigen Demand (BOD and COD) values, pathogen bacteriological (E-coli) loads and occurrence of helmints, etc. Criteria applied for the different purposes and technologies of irrigation can be found in guidelines of the Food and Agricultural Organisation of the UN (FAO 1992).

Water releases from reservoirs may be required to meet specific ecological criteria. This is more common in systems that involve the use of river carriers to convey water between the reservoir and the irrigation project. The flow regime may involve additional releases that are often in conflict with the irrigation demand. While the operation of the irrigation distribution system may not be directly affected, the allocation of water for environmental maintenance may result in restrictions being imposed on the security of supply to users.

4.3.2 Determining Factors for Drainage Services

In most irrigation areas the drainage service consists of providing an outlet for the farm drainage effluent. However, in some areas, especially in deltas, the drainage service also consists of regulating maximum and minimum water level to control the level of the ground water. Drainage services should therefore be related to their specific purpose, which is determined by the source of excess water, groundwater tables, land use, soils and the hydro-topography[9] in the drainage service area. Other factors to be included are climate, precipitation patterns, hydraulic infrastructure and environmental regulations.

Drainage Objective

Control and removal of excess water is as necessary for plant growth as irrigation. Excess water from irrigation, rainfall or a combination of both often hampers agricultural productivity. The removal of excess water from the land surface and excess water and salts from the soil by artificial means is called *land drainage*. In agriculture its purpose is to make the land more suitable for agriculture and is aimed to increase production, to sustain yields, or to reduce production cost – to maximise the net return of the farming enterprise (ILRI, 1979).

A distinction must be made between the drainage of water on the soil surface, in the root zone and of groundwater. In absence of shallow groundwater tables, surface and subsurface drainage problems may be due to high precipitation or irrigation intensity and unfavourable soil structure. These problems usually can be overcome either by installing surface drainage or by improving the soil condition. For this purpose agricultural drainage services are provided to enable farmers to dispose of the excess water from their agricultural lands. This service consists of securing the disposal of farm effluents by providing access to the main drainage system through a drainage outlet.

---

[9]. Hydro-topography is the relative elevation of land compared with the hydrologic environment. (Suryadi, 1996). Hydro-topography determines the ability of an area to drain by gravitation which is of special importance in river deltas, polders, tidal lowlands and floodplains of rivers.

Where high ground water tables are present, drainage can be achieved by lowering the groundwater through a subsurface drainage system. In this situation, the objective of the drainage service is to provide an outlet for subsurface drains to enable the farmers to control the groundwater level in their fields.

The provision of services could further involve maintaining minimum ground water depths within certain margins of fluctuation to:

- control water supply to the crop through capillary rise, which can be important in rainfed agriculture;
- control chemical processes in soils like peat and acid sulphate soils where low groundwater tables will induce oxidation processes. These will result in land subsidence for peat and development of acidity and toxicity in acid sulphate soils;
- minimise subsidence due to drainage of soils;
- prevent damage to foundations of buildings and structures.

In addition to agriculture, drainage systems are also used for removal of excess water from non-agricultural areas like nature, urban and industrial areas. Besides, drainage systems are also used for groundwater level control and as carriers to dispose domestic and industrial wastewater. According to these functions, drainage systems can be assigned specific purposes, each of them with their own service specifications and conditions and requirements for the physical and management infrastructure.

Drainage of excess water due to precipitation or irrigation on the surface and in the groundwater within the drainage area is referred to as internal drainage. Another drainage service often provided is external drainage. Its purpose is to convey "outside" water either through or bypassing the area under management. This external drainage function relates usually to rivers or streams inside or at the boundary of the managed area with their drainage basin located outside this area. Often external drainage facilities go in parallel with provisions for flood mitigation or flood protection. As the occurrence of floods is also of stochastic nature, the service to be provided will therefore be expressed in a degree of safety related to the occurrence of an extreme floodwater level or discharge event.

Climate and Precipitation Regime

Climatic characteristics are an important factor in drainage service provision. In arid and semi-arid zones where rainfall is insufficient to leach salts from the soil, drainage focuses on the control of waterlogging and salinisation. In the humid tropics, drainage focuses on mitigating water logging and preventing flooding. The intensity, duration and distribution of rainfall are of stochastic nature. The amount of surface runoff and consequent changes in surface and groundwater levels depend on the topography and the soil characteristics and the intensity and duration of the storms. Therefore service levels must be related to either the probability of occurrence or exceedance of a given rainfall event or water level.

Many areas in the humid tropics are subject to monsoonal precipitation requiring extensive drainage facilities to evacuate the excess water. Under these conditions, the provision of service is concerned primarily with the degree of protection offered to different parts of the system on the basis of their topographic location. Protection of low-lying areas is more costly than that of high areas. Often, this type of drainage must rely on pumping to evacuate the excess water such as in the Netherlands, China and Vietnam. The required capacity of the system must then be related to the features of the land and the rainfall characteristics of the region.

Drainage Policy

The formulation of a drainage policy is concerned with the level of protection to be provided to areas with different production capacity. Consideration must be given to the cost and the benefits that can be obtained from protecting the land. Sometimes, drainage policies are part of regional, rural or urban development plans. Frequently, drainage policy in the tropics is aimed at providing protection to all areas subject to flooding regardless of the drainage intensity required. This approach is often justified only on the basis of land use for production of low value crops such as rice. It may not be economically feasible to drain such low-lying areas although often the social situation locally demands such protection, especially if there is urban development. In the absence of urban or industrial development, these areas could be put to other alternative uses such as fisheries that would allow the excess water to be used as a resource rather than being removed.

The authors believe that more effort must be made to develop integrated drainage policies that contemplate alternative land uses. The introduction of drainage fees might be a catalyst to promote more rational drainage policies. The drainage service specifications must involve extensive consultation with the users based on an agreed level of service and cost of drainage provision. This may be differentiated according to areas with different land use and capability.

Land Use

Land use is a determining factor in drainage service requirements. A distinction has to be made between agricultural land, un-built areas and built-up urban areas.

Firstly, the peak runoff of built-up areas will be relatively high because of the impermeable surface and the shorter time of concentration. In un-built areas the peak discharge will be relatively low due to infiltration and temporary storage in local depressions.

Secondly, the degree of protection adopted against inundation is related to the value of the property or crops and the damage that will occur if the standard rainfall or water levels are exceeded. The level of protection for agriculture is usually lower than that for urban areas and industrial complexes. Even within agricultural areas different levels of protection can be desired as some high value crops may be more sensitive to waterlogging than others that have a lower economic value. Hence, there can be a direct relation between the level of protection adopted, the value of protected property and the cost of service provision. A higher level of protection entails a larger capacity of the conveyance and storage system, which translates directly into higher infrastructure and operating costs.

Crops and Soils

The sensitivity of crops to waterlogging and salinity is the main factor for the design of agricultural drainage services. Crop sensitivity determines the maximum allowable depth and duration of inundation and the occurrence of exceedance of maximum allowable groundwater levels. The required on-farm groundwater levels determine the required water level in the collector and main drains. Lifting may be necessary when target levels cannot be achieved by gravity. These target levels have to be translated in water levels in the main drainage system controlling the groundwater levels. These criteria differ for each crop type and crop stage. An agreement with the users must be obtained on the level of service to be provided, taking into consideration the temporal and spatial variations in cropping patterns. More flexibility in drainage service will certainly involve a higher cost of service. The practicality of a flexible drainage service depends on the scale of variation in crops and the size of the landholdings. It is easier to accommodate the desires of large landholders with a single crop type than of smallholders with a wide range of different crops.

Hydraulic conductivity and storage capacity or specific yield are the most important soil characteristics for drainage. They determine the drain spacing of field drains and it has an effect on the peak discharge of subsurface drains. The capacity of the collector drains must be appropriate to accommodate these peak discharges from the farm systems in order to meet the service specifications for water level variations and disposal capacity.

Hydraulic Infrastructure

Drainage service criteria determine the requirements for hydraulic infrastructure in the form of drainage density, conveyance capacity, storage capacity and water levels. In cases where gravity disposal is not available or too costly pump capacity must also be included.

In many instances, drainage services have not been developed because of the availability of a natural drainage network. However, drainage canals for evacuation of surface water have been provided in most irrigation schemes built in recent decades. The development of subsurface drainage systems is however in many countries still in its early stages.

Storage capacity plays an important role in the design of drainage canals and pumping stations. A larger storage or retention capacity within the canal system lowers the requirement for conveyance and pumping capacity to achieve the same service objective. There is of course an optimum storage capacity that is related to the value of the land and the cost of storage facilities.

Gravity drainage systems usually require very little operation. The set point for water level control structures can be fixed for prolonged periods, often on a seasonal basis e.g. dry and wet season, summer and winter, etc. These systems are of course subject to fluctuation of surface water levels and groundwater levels. The same applies for pumped schemes. The pumps must be started when a certain water level is exceeded and shut down again when the water level drops below a certain level. This process of local control can be automated and if many pumping stations or water level regulators are centralised control can be considered.

A particular situation arises with the disposal of water through tidal outlets such as in the Mekong Delta in Vietnam or in the tidal lowlands in Indonesia. Many irrigation schemes are located in deltas where tide levels influence the drainage capability. Storage capacity is important in these systems to overcome the period that no water can be disposed by gravity due to high tide. This is an important variable in the determination of the service levels and especially the service cost of the system. Sometimes a combination of gravity disposal and pumping is required while in other cases dual purpose pumping facilities are used for drainage and irrigation water supply.

Environmental Regulations

Drainage services must often comply with environmental regulations concerning the disposal of drainage effluents. These are primarily aimed at protecting the quality of the receiving waters. The main pollutants in drainage water are excessive salts, nutrients and pesticide residues from farms and animal husbandry. Monitoring the quality of drainage effluents can take place at the farm outlet or other levels in the drainage collection network (Box 4.2). Other pollutants that can occur in drainage water are high acidity and toxicity levels due to leaching of acid sulphate soils or peat soils. Non-agriculture outlets like sewage, treatment plants and industrial waste can be monitored on an individual basis. Measures can be imposed to meet effluent standards such as treatment of wastewater, alternative waste practices (especially manure from animal husbandry) and waste-accounting (e.g. salt balances). Target water levels may be imposed to reduce oxidation of peat and acid-sulphate soils.

The delivery of drainage services in the humid tropics must be able to cope with the stochastic nature of rainfall, especially in areas where rainfall is the dominant factor. This is in contrast with irrigation service, which is of a deterministic nature and usually relies on a single purpose system used to supply water to farmers. Other factors further distinguish the provision of drainage services from that of irrigation services. In addition to farmers, drainage systems serve other water and land users including urban, industrial and other non-agricultural land uses. Thus drainage may involve a number of different customers with different interests and different level of service desire. They all should be consulted in the customisation of drainage service specifications.

Box 4.2 Some water quality norms for surface waters (Delfland, 1993).

Parameter	Maximum value	
General	Water free from visible pollution and smell	
Temperature	$< 25\ ^{\circ}C$	
O$_2$	5 mg/l	
pH	6.5 - 9	
Nutrients		
P	0.15 mg/l	
N	2,2 mg/l	
Chlorofyl	100µg/l	
Ammoniak	0.02 mg/l	
Salts		
Chlorides	200 mg Cl/l	
Fluor	15 mg F/l	
Bromide	8 mg Br/l	
Sulphate	100 mg SO$_4$/l	
Bacteriologic		
Thermotolerant coli	20 MPN/ml	
Metals (µG/L)	Target	Max.
Cadmium	0.05	0.2
Mercury	0.02	0.03
Copper	3	3
Nickel	9	10
Lead	4	25
Sinc	9	30
Chromium	5	20
Arsenicum	5	10

As the range of interests connected to the drainage system expands, more factors must be considered in the process of service customisation.

4.4 FORMULATION OF LEVEL OF SERVICE SPECIFICATIONS

4.4.1 Desired Qualities of Level of Service

From the perspective of the users, there are several elements that determine the quality of the water supply and drainage service including flexibility, reliability, equity and adequacy. Reliable, equitable and predictable water supply and reliable drainage services are preconditions for good water management on the farm or below the supply outlet.

Adequacy

Adequacy of irrigation service is a measure of the ability of the water supply schedule to meet the water demand for optimal plant growth. It is often expressed as the ratio of the amount supplied to that required by the crop. However, a more appropriate definition of adequacy should take into account the users and farming needs to accommodate irrigation with other farming operations such as land preparation, cultural operations during the crop cycle, etc. The adequacy of supply is a function of the area of land irrigated, crop consumptive use, irrigation losses and salt leaching requirements and the ability of the irrigation delivery system to meet these requirements. The level of service selected will determine how well these requirements are met both in time and space within the irrigation scheme.

Drainage adequacy is a measure of the capability of the drainage system to cope with excess water situations to minimise or prevent associated damage. Usually storm runoff will be the determining factor for drainage capability. Adequacy of drainage services can therefore be expressed in the capability of a drainage system to evacuate a standard design storm.

Reliability

Reliability is a measure of the confidence in the irrigation system to deliver water and in the drainage system to evacuate excess water and/or control (ground-) water levels as specified by the level of service. It is defined as the temporal uniformity of the ratio of the amount of water supplied to the required or scheduled supply.

Reliability of water supply is important to users because it allows the proper scheduling of farming operations to achieve optimal crop yields. It is possible that a water supply may be reliable but inadequate or vice-versa. Farmers may prefer high reliability of supply to greater adequacy because the level of security of good crop yield is higher albeit with reduced areas. Reliability is also a function of the form of water supply. In systems that supply continuous flow, reliability refers to the expectation that a certain discharge or water level will be met or exceeded. Under these circumstances variability is the main concern. In systems that supply water intermittently, e.g. rotational irrigation, predictability of the timing of supply is the main concern.

The importance of reliable drainage services for agriculture very much depends on the climatic and topographic situation of the area concerned. In arid and semi-arid zones the need for adequate surface and sub-surface drainage systems may be ignored for a long time as the impact of such system is more of a preventive nature. Users only realise the need for such a system when salinity and water logging conditions are already present. In humid tropics and in flat areas, surface drainage is considered more important as inundation will occur frequently if proper drainage facilities are lacking. In these cases the reliability of drainage services is more important.

Equity

Equity in irrigation is a measure of the access to a fair share of the water resource according to the amount specified by the water entitlement. It is difficult to provide a general measure of equity given that wide variations in water right systems may exist. However, in general it can be defined as the actual supply of water to users in relation to the allocated share. The term is often confused with equality. However, equality denotes a condition where all farmers should be made to stand equal in the face of either abundance or shortage of water and regardless of their size (big or small) or location (head or tail).

Equity in drainage reflects the spatial and temporal spread of the duration of waterlogging and inundation and the consequent damage. Waterlogging will usually occur most in depressions that also suffer most in case of inundation. Equity in drainage therefore means a sharing of risk by for example equal limiting measures of drainage outflow from farms or tertiary units throughout the area. This would result in identical water excess in identical rainfall events. The outflow capacity might be related to value of crops or assets on the land to be drained that will also be reflected in the cost and price of providing such higher level of service.

Inequity in irrigation and drainage often go together. Besides irrigation water shortages, flooding and waterlogging also occur most often in the tail end areas of irrigation systems as these are located lowest in the landscape.

Flexibility

The ability of users to choose the frequency, rate and duration with which irrigation water is supplied determines the degree of flexibility of supply. In general, a higher level of service entails a greater degree of supply flexibility. Rigid schedules can be delivered with simpler hydraulic control infrastructure and therefore lower capital and operating costs. The main drawback of rigid systems is their inability to supply water according to the needs of crops or irrigators and to take advantage of rainfall as part of the water supply.

More flexible schedules require the implementation of more sophisticated hydraulic control infrastructure and operation procedures. They also require conveyance systems with greater capacity to meet the possibility of concentrated demand by the users. These requirements will often translate in greater capital and operation costs. However, a more objective evaluation of the delivery schedule should also take into account the benefits that can accrue from more flexible irrigation supply such as higher productivity and quality of the agriculture production. In fact, it is often claimed that a flexible irrigation service can stimulate farmers' investment in better agricultural technology for crop production leading to higher farmers' income.

Flexibility in water application at the farm level can also be obtained by making use of field reservoirs or conjunctive use of groundwater. In both cases the delivery schedule to the farm outlet can be rigid in which case flexibility can be enhanced and managed by individual farmers or groups of farmers.

The degree of flexibility of drainage services is determined by the ability of users to choose the time, rate and quality with which excess water can be disposed of. Also for drainage a higher level of service entails a greater degree of flexibility of disposal.

Table 4.1 Desired qualities of level of irrigation and drainage services

Service quality	Irrigation	Drainage
Adequacy	Ability to meet water demand for optimum plant growth	Ability to dispose excess water in minimal time to prevent damage.
Reliability	Confidence in supply of water	Confidence in ability to dispose excess water
Equity	Fair distribution of share of water shortage risks	Fair distribution of inundation risks
Flexibility	Ability to choose the frequency, rate and duration of supply	Ability to choose the time, rate and duration of disposal

4.4.2   New or Rehabilitated Irrigation and Drainage Schemes

The development of a new irrigation and drainage scheme or the rehabilitation and modernisation of an existing scheme must be planned to meet a set of level of service specifications agreed upon with the users. The planning of new schemes provides more flexibility for the agency and users to tailor the appropriate hydraulic control infrastructure and the required management inputs to arrive at an agreed level of service and sustain it after commissioning the new scheme. The main conditioning factors that must be taken into consideration during this process are:

- water rights governing the access to water by farmers and agency;
- Government agricultural, environmental and irrigation and drainage policies;
- soils-crops-climate interactions;
- characteristics of the surface and ground water resource.

Figure 4.2 illustrates the interaction between the level of service determining factors. A set of level of service options is developed based on the desired level of service expressed by the users. In this process, a firm link is established between the level of service specifications and the associated cost of service. At the core of this process is the evaluation of the various types of hydraulic control options and associated management that is required for each level of service. In addition to hydraulic criteria, the selection of flow control technology must be based on the management arrangements and capacity that is required to operate the technology. A more detailed discussion on the relation between level of service and hydraulic control technology is provided in Chapter 5.

The management arrangements required to deliver the agreed level of service must be put in place before the commissioning of the new infrastructure. The management and operation requirements for implementing and sustaining the agreed level of service are usually very different from those that have been in place during the development stage. Of key importance is the establishment of an on-going training program to support the

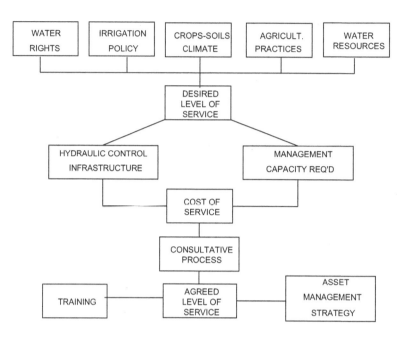

Figure 4.2 Formulation of level of service specifications for new or rehabilitated irrigation and drainage schemes

strategic management plan. Training which is often implemented on an ad-hoc basis as part of development projects and usually not related to the achievement of overall management goals by the irrigation agency is not effective.

4.4.3   Existing Irrigation and Drainage Schemes

The formulation of level of service specifications for many existing irrigation and drainage schemes often does not involve a proper consultation process with the users. In some cases the level of service is dictated by government policy or unilaterally formulated by the irrigation and drainage authority. Yet in many other cases no level of service criteria is formulated. Many irrigation schemes have been developed based on standard principles and designs that are applied without questioning the service the irrigation or drainage system is expected to provide. Furthermore, the operation and management aspects of the system are usually considered only after commissioning, unnecessarily limiting the service options.

In formulating the level of service specifications for an existing scheme, the hydraulic control infrastructure imposes constraints on the range of options that can be evaluated as part of the consultative process. In such cases, changes in management can play a significant role in shaping the level of service. Figure 4.3 describes these interactions between factors that influence the level of service. In addition to the conditioning factors for new systems, the type and condition of the existing hydraulic control infrastructure and its management must be considered. This level of service may be satisfactory or fall short of the level of service desired by the users.

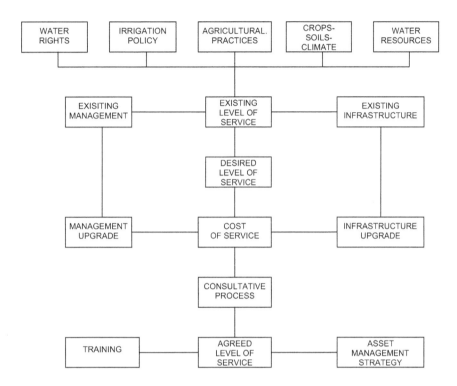

Figure 4.3 Formulation of level of service specifications for existing irrigation and drainage schemes

Users through the consultation process can express their desired level of service. This is often a higher level of service than that being currently delivered to them. The provision of the desired level of service may require both an upgrading of the system infrastructure and of agency's management capacity. A comprehensive asset management program must be formulated to evaluate the maintenance, replacement and upgrading of the hydraulic infrastructure required to meet the desired level of service and enable management to determine the cost associated with the higher level of service. A detail discussion on the formulation of asset management programs is provided in Chapter 6. During this process, several upgrading options are formulated and evaluated with the involvement of the users and a selection is made. The delivery of the agreed level of service normally involves changes in the management arrangements. These changes must be reflected in the strategic plan of the agency. Training of agency staff in the operation of the system plays a pivotal role in ensuring an effective delivery of the agreed level of service.

4.5 LEVEL OF SERVICE SPECIFICATIONS

Service oriented irrigation and drainage is based on establishing a set of specifications that will govern the operation of the system. These specifications serve two purposes: (a) provide a set of rules against which the operational performance of the system can be measured, and (b) provide a set of rules that govern the conditions for service delivery. The first set of specifications must constitute clear, quantifiable and measurable operational parameters that permit the comparison of the actual operation against the set management targets. The second set of rules refers to the provisions under which the service is provided. This set of provisions is system-specific and includes aspects of user and agency accountability and specific service restrictions.

4.5.1 Level of Service Specifications for Irrigation

The typical set of operational specifications for irrigation services includes the following:

- *rate, duration and frequency of supply*: These are the main parameters that define the irrigation supply flexibility and adequacy. There is a wide range of irrigation supply schedules that can be adopted for the provision of service. These can range from very rigid supply such as fixed rotation to very flexible supply schedule such as on-demand systems. Chapter 5 will analyse various supply schedules in more detail;

- *height of supply:* The height at which water is supplied to the farm is critical to the ability of implementing adequate design and water management practises. The height of supply is determined by the maximum operating levels of the supply channels. In addition to specifying the height of the water level, level of service specifications may include an acceptable level of fluctuations;

- *supply pressure:* Pressurised irrigation such as sprinkle and trickle systems require that the operating pressure be maintained within a certain range for which they were designed. Supply pressure outside this range will result in lower efficiency and uniformity of application. The desired pressure range together with the amount and frequency of pressure fluctuation must be part of the level of service specifications;

- *supply monitoring.* Compliance with the stipulated level of service specifications requires effective field monitoring procedures. The required arrangements vary according to the degree of automation of the hydraulic control infrastructure. Fully automated systems allow the remote monitoring of water level and discharge in the canal systems. Most commonly, this task must be performed manually by system operatives. The level of service specifications must make reference to the location and frequency with which the operational parameters e.g. water level, discharge, quality; etc, will be observed;

- *security of supply*: This refers to the long-term expectation of supply. It consists of a series of supply availability values and their associated probability of occurrence. These values are derived from the system reliability analysis and are often associated to the allocation priority and restrictions in supply in case of water shortages;

- *supply water quality*: This is the criterion related to the upper limits of pollutants in the irrigation water including, nutrients, and pesticides that may be deleterious to humans and animals. The threshold values selected are usually related to the use(s) made of the water supply.

This is not intended to be an exhaustive list of service specifications. As indicated earlier, the site-specific nature of irrigation often requires specific rules to be developed, appropriate for the nature of system. These can vary widely between systems, reflecting the system characteristics (Box 4.3).

The main conditions that apply to service delivery are:

+ *payment for supply*: The basis for water charges can vary widely from a flat rate per unit area to a volumetric rate. The tariff structure has the potential to promote a more efficient use of the resource. A flat rate per unit area is usually associated with a fixed allocation of water and rigid delivery. In these cases tariff structures may include a diversification to the type of crops irrigated. In many schemes where this price structure is used the water property and water use rights are linked to the land rights. Volumetric charges can better reflect the value of water as an economic good since charges are proportional to the volume of resource used. By charging on a volume basis, it is also possible to develop a differential tariff structure in which successive segments of the total allocation are charged at a different rate;

+ *points of supply*: Irrigation water may be supplied to individual users or to users groups. It is necessary to specify the maximum number of points or outlets permitted for each discrete and separate land unit. For simplicity of operation, the number of supply points is usually restricted to a single point, e.g. the head gate of a farm unit or the tertiary unit offtake;

+ *water ordering*: In schemes with a more flexible water supply arrangement, the individual users or user groups are able to request water according to their need. A clear set of rules must be established for the user to request water deliveries and for the irrigation agency to carry out the delivery. Depending on the type of hydraulic control and management, there is often a certain amount of advance notice that the user must provide before water can be supplied. The interface level where the delivery takes place varies according to the particular user-operator arrangement for the scheme. In schemes with increased user responsibility for management of the irrigation infrastructure, users may receive the delivery at the inlet of secondary canals. In such cases, additional specific arrangements may be necessary to operate the lower reaches of the system;

+ *supply restrictions*: The imposition of restrictions on supply during times of low inflows may be necessary to conserve supplies. The restriction policy may differentiate between different user groups so that the severity of restrictions on supply to different users reflects their ability to cope with shortfalls in supply. The supply restrictions are usually stated as a series of "trigger levels" at which restrictions are applied. In regulated systems, reservoir rule curves are used to determine the trigger levels;

+ *allocation priority*: Water supply priorities may differentiate between different crops and different water uses. Some uses such as town water supplies, industrial supply, supply for perennial crops (orchards) etc, may be assigned higher priority of use. In this instance, other uses can only be supplied after higher priority demands have been met;

+ *service provision during the year*: Irrigation supply must be interrupted to carry out normal maintenance and repairs at certain times during the year. Moreover, the beginning and end of the irrigation season may be determined by weather conditions. The criteria to determine the regular commencement and end of irrigation supplies must be clearly stated as part of the service agreement. Sometimes, temporary interruptions are also caused by emergency or system failures.

The flexibility, adequacy, reliability and equity of water supply and the ability of the organisation to meet the established service specifications determine the quality of service (Box 4.4). No single quality of service criteria is sufficient to rate the quality of service. In principle, unconstrained on-demand systems can provide the greatest flexibility of water supply. However, reliability of service is crucial to the irrigators' confidence in the ability of the organisation to deliver the specified service. In all cases, an effective monitoring and evaluation system is required to assess the performance of service delivery. Chapter 7 will discuss performance monitoring in more detail.

4.5.2 Level of Service Specifications for Drainage

There is not a single typical set of service specifications for drainage. The variation in drainage functions and the diversity of drainage requirements for different situations require site specific specifications. Some basic generic specifications are provided below from which more specific specifications can be derived; although this is not exhaustive list and may need to be expanded to accommodate local needs.

- *rate, time and duration of disposal*: The capacity of the drainage system is designed to accommodate disposal of excess runoff for a standard precipitation event of certain duration and intensity. The period of disposal is determined by the allowed duration of inundation. Storms exceeding the design criteria will result in an increased disposal discharge from the upstream part of the system resulting in reduced possibilities for disposal downstream. If in such case discharges from upstream fields can be restricted then inundation periods along the drainage area can be distributed more equally. Discharge restrictions are imposed to provide equitable service to the upstream and downstream parts of the system. Restrictions on time and duration of disposal are usually imposed on drainage schemes that depend on tidal outlets or pumping outlets. Less restrictive time and duration of disposal leads to a higher level of service. Such level of service would also imply a higher cost of service ;

- *maximum and minimum water level*: proper functioning of the subsurface drainage system requires a water level at the disposal site lower than the field drain outlets. Maximum target water levels (with a probability of exceedance in case the same drains are used for surface runoff) need to be specified in conjunction with the field drainage systems. Minimum water levels need to be specified in case prevention of subsidence and oxidation is necessary and of crops making use of capillary rise from the groundwater table. These minimum levels in turn relate to controlling water level in higher order drains;

- *reliability of disposal*: The capacity of the drainage system depends on the design capacity and on the state of maintenance. Deferred maintenance can result in unnecessary prolonged inundation if the design storm occurs. However, deferred maintenance will remain undetected until a large storm occurs. Hence, the maintenance level has a direct relation to security of disposal and must be stated explicitly in the formulation of the system reliability;

♦ *flood protection level:* Safety against inundation from external drainage sources must be provided on a whole area basis rather than for individual land holdings. Flood protection levels are usually expressed in terms of occurrence of flood levels. Hence, safety measures will be expressed in height of flood embankments required to protect against a certain flood level with a certain probability of occurrence.

A typical set of conditions for drainage services include:

♦ *cost of disposal:* The cost of the drainage service includes the infrastructure for provision of drainage within the drained area such as collection, conveyance and disposal, possible treatment plants and disposal sites like evaporation basins and flood control or protection works for external flooding. The recovery of these costs depends on the service charging policy. The tariff structure can be a flat rate based on the size of drainage area or it can be further developed by including the size of drainage outlets, the value of crop or property protected, the degree of pollution, or the security levels provided;

♦ *point of disposal:* The point of disposal must be fixed if use of drainage services is conditional and restrictions apply on disposal rates and quality of the disposed effluent. Usually disposal points serve an entire drainage catchment including a group of farmers or defined urban or industrial areas. To facilitate better monitoring and control, individual or specific outlets may be assigned to users with a higher probability of polluted effluents. These points of disposal will be explicitly mentioned in the service agreement;

♦ *disposal restrictions:* Disposal of drainage water can be restricted if water is contaminated or if the capacity of the receiving carrier is reached or exceeded. The drainage effluent may contain a range of pollutants including high salt content, nutrients from fertiliser applications and animal husbandry and pesticides residues. Environmental regulations may set out the water quality standards that must be complied with by farmers and agency providing the drainage service. These standards are usually enforced by an external government organisation, or may be enforced by the drainage organisation on behalf of government. Environmental regulations may be enforced by a separate agencies which operate at different levels in the system, eg. the drainage authority monitors drainage effluents from the individual properties and it must in turn comply with effluent standards at the system outlet imposed by a different agency;

♦ *disposal priorities:* In case of restricted disposal, priorities of disposal may be given to strategic and economic areas like cities, industrial complexes, horticulture areas, etc. In such cases, other areas may only dispose of their excess water after the priority areas have been drained to a predetermined level;

♦ *service provision during the year:* A drainage system must be functional throughout the year. However, in exceptional cases major maintenance or renewal works might be required that will restrict or interrupt the service. Maintenance and renewal works should therefore be planned in dry seasons when the probability of occurrence of big storms is minimal or in case of subsurface drainage during the off cropping season.

4.6 SERVICE AGREEMENTS AND ACCOUNTABILITY MECHANISMS

The provision of irrigation and drainage services according to a set of specifications requires a set of "working rules" that delineate the procedures and mechanisms to govern the relation between agency and customers (users). Such rules forbid, permit, or require certain actions to be acted upon. Central to the concept of working rules is that they must be enforceable. Ostrom E. (1990) states *"...working rules [must be] common knowledge and [be] monitored and enforced. Common knowledge implies that every participant knows the rules, knows that others know the rules, and knows that they also know that the participants know the rules."* Therefore, service relations and the services provided need to be clearly defined and transparent administrative procedures must be put in place supported by effective accountability mechanisms. These arrangements are formalised through individual contracts between the service provider and customers or through the customers service agreements.

4.6.1 The Service Agreement

It is necessary for all service relationships to define services (transactions) and the conditions attached to them, and payment required for obtaining these services. These must be stated in quantifiable and measurable terms that are easily monitored and controlled. These can be formulated in service agreements in the form of contracts that contain details on the level of service to be provided by the organisation, the obligations of customers and the organisation and the process for resolution of conflict should these arise.

Service agreements consist of two main elements: (a) Transactions and (b) Accountability mechanism. The *transactions* deal with the nature of the service provided in accordance with the irrigation supply or drainage specifications and the payment and other obligations from customers.

The *accountability mechanisms* are to ensure that obligations from the service provider and the customers are met and that the provision of services and its payment arrangements can be adjusted according agreed procedures. Snellen (1996) identifies six items that should be specified in a service agreement:

- specification of services that will be provided;
- amount and form of payment of service by users;
- monitoring procedures to verify whether services are provided as agreed;
- liabilities to both parties for not fulfilling the agreement;
- relevant authority that will settle conflicts;
- process for reviewing and updating the service agreement.

4.6.2   Accountability Mechanisms

In the context of irrigation and drainage organisations, three domains of accountability can be identified that will result in three working rules[10] levels: operational accountability, strategic accountability, and constitutional accountability (van Hofwegen, 1996).

Operational accountability involves the mechanisms for monitoring, evaluating, controlling and enforcing the implementation of the service agreement. It consists of all the monitoring procedures required to verify the delivery of the level of service by the agency and the users' adherence to the delivery conditions and payment of service. The core of the operational accountability is the formulation of a service agreement between the irrigation and drainage organisation and the clients (users). A key element for the implementation of service agreements is the transparency of the process involved in both the delivery of service and the obligations of both the agency and their clients (Box 4.7).

Strategic accountability is related to the mechanism that users have to control the formulation of the service agreement. It is necessary to have clear rules for consultation between the agency and users in which water users or their representatives participate in the process of discussing and approving proposals prepared by the irrigation and drainage organisation.

Constitutional accountability relates to the mechanisms by which users can influence the strategic decision making process of the organisation. Users can influence this process in various ways - depending on the political context - through the election of their political representatives, through the minister responsible for the portfolio and through their representatives on the organisation's governing body.

4.6.3   Agreed and Declared Levels of Service

The level of service has been defined earlier as a set of operational standards set by the irrigation agency in consultation with irrigators, other affected parties and the government to manage an irrigation and/or drainage scheme. This presupposes a consultative process, which involves government, users and agency to establish an agreed level of service, procedures involved in establishing a level of service and the price of service. The willingness and capacity of users to pay for the services depends on the profitability of their farming enterprises, the reliability of services and the severity of penalties imposed for lack of payment.

The provision of a certain level of service is greatly influenced by the physical environment (availability of water and land resources), management environment (socio-economic, institutional and legal frameworks) and the hydraulic infrastructure. The establishment of the level of service is also part of a development process, which requires a continuous process of consultation, adjustment and adaptation within an effective accountability system. In this regard, a differentiation must be made between an *agreed level of service* which is demand driven, and a *declared level of service* which is supply

---

[10] Ostrom identifies three levels of nested working rules: (a) *operational rules* that affect the day-to-day decisions made by users concerning when, where and how to withdraw water; (b) *collective-choice rules* that are used by appropriators, or external authorities in making policies and operational rules and (c) *constitutional-choice rules* that determine who is eligible and the specific set of rules to be used in crafting the set of collective-choice rules

driven. The difference relates to the degree of consultation between the organisation and customers during the process of defining the level of service. This often includes also the level of cost to be recovered from the beneficiaries and the level of government subsidies. Customers (users) can thus have a different degree of intervention in the formulation of the level of service ranging from a varying degree of customer participation to services dictated only by the managing authority.

In the first case we speak of service oriented management in which the provision of services is based on *explicitly agreed level of service and agreed level of cost.* This agreed level of service is the result of a consultation process in which all the interested parties (government, agency and customers) are involved. The level of service to be provided is balanced against the cost associated with the delivery of that particular level of service and the objectives and targets for service delivery are formulated in a service agreement. In service oriented management the needs of the customers are of primary importance. The service provider adjusts its management practices to meet the needs expressed in the consultation process. Moreover, the performance of the managing organisation is measured against the fulfilment of this agreed level of service.

In the second case we speak of performance oriented management based on a *declared level of service to be provided at a declared level of cost.* This means that responding to the needs of the clients is a lesser priority. The customers adjust their agricultural practices to the services provided. This is usually the case in situations where resources are scarce and irrigation is considered an activity necessary to meet social needs. Nevertheless, the principle of balancing the level of service against the cost of service provision still remains. In this situation, often the price the users pay for the service is a fraction of the cost of providing this service. The remaining part of the cost to provide the service must be covered by government subsidies. A clear statement of level of service and accountability mechanisms is as important in systems with declared levels of service as they are in systems with agreed levels of service to achieve high levels of performance. This is often lacking in publicly managed systems resulting in a low level of recovery of water charges and inadequate levels of subsidies. As a result, there exist poor accountability between the managing organisation and customers. This coupled with inadequate levels of expenditure in infrastructure maintenance and renewal often results in low levels of service and lack of sustainability of the irrigation and drainage infrastructure.

Box 4.3. Comparison of Level of Service Specifications for 4 Irrigation Schemes

SERVICE SPECIFICATIONS	Societe du Canal de Provence France	Goulburn-Murray Irrigation District Australia	Triffa Irrigation Scheme ORMVA de la Molouya Morocco	Acequia Real del Jucar Spain
Type of Organisation	Public Corporation, shares owned by local government, banks and chambre of agriculture	Public Corporation, infrastructure owned by government	Public Corporation, infrastructure owned by government	Autonomous Irrigators' Association
Operation Concept	On-demand: Unconstrained	On-request: Unconstrained	On-Request: Constrained	Imposed
Flow Rate	Unconstrained upto maximum	Constrained only by channel capacity	Constant: 20 , 30 or 40 l/s	Fixed
Frequency	Unconstrained	Unconstrained with 4-day notice.	Constrained: number of deliveries related to availability of water	Fixed rotation. Interval based on crops, soils and climate.
Duration	Unconstrained	Unconstrained	Constrained: maximum duration based on crop and flow rate	Fixed, proportional to land water right
Heigh of Supply	Canal levels and pressure levels	Channel design level	Channel design level	N/A
Operation Monitoring	Deliveries monitored through double counters/flow meters (agency and user), Monthly readings	Agency to ensure ordered flow rate, provided customers adhere to scheduled start and finish times.	Farmers sign receipt after delivery	N/A
Delivery Performance	According service contract. Different contracts for different uses. Target 96% of time low pressure delivered unless stated otherwise	Target: 86% of orders delivered on day requested,	Target: delivery in full accordance with agreed irrigation schedule.	N/A

Box 4-3. Continued

Condition	Societe du Canal de Provence France	Goulburn-Murray Irrigation District Australia	Triffa Irrigation Scheme ORMVA de la Moulouya Morocco	Acequia Real de Jucar Spain
Water Charges	Fixed + Volumetric Fixed charge based on subscription of delivery rate, Volumetric based on volume delivered. Gross average US$0.10/m3 full cost recovery including asset renewal	Volumetric: US$0.021/m³ Full cost recovery including asset renewal annuity	Volumetric: price varies for gravity, lifted and pressurised water from US$0.020/m² – US$0.040/m² Government subsidy to cover cost recovery deficit	? N/A
Points of Supply	One point of delivery per contract holder (different uses)	One supply point per property fitted with door and meter wheel	One supply point per group of farmers, farmers rotate supply.	One supply point per group of farmers, farmers rotate supply.
Water Ordering	On demand so no ordering	Telephone ordering system – 4-day notice required	ORMVAM announces an irrigation cycle, farmers can request time and duration of delivery. Schedules are drawn up and agreed upon.	N/A
Supply Restrictions	In case of water shortage a system of water orders is introduced and allocations are made in proportion to these orders	Supply allocated equitably to all customers when demand exceeds supply	Prior to season restriction announcements for crop types are made. During season equitable distribution between permitted crops.	Rationing with priority given to perennial crops
Water Rights	According contracts	Transferable permanently or temporarily	Attached to landownership.	Non-transferable, attached to land.

Box 4.4 Accountability mechanisms within the Triffa Irrigation Scheme, Morocco and the Rijnland Waterboard, The Netherlands

Triffa Irrigation Scheme, Morocco

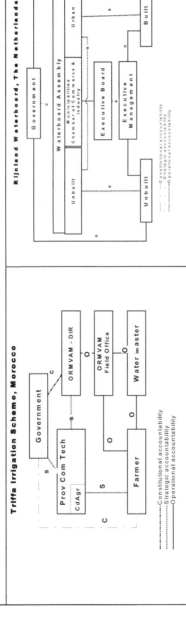

Operational accountability: Direct relation between the farmers and the ORMVAM field officers for scheduling and delivery. Farmers sign a receipt for each delivery of water. Sworn officials with powers to impose fines handle water thefts and other disputes. Third time repeat offenders will be charged and the case tried by a legal tribunal. ORMVAM can discontinue water supply service due to default in payments of water charges.

Strategic accountability: Coordination with farmers and other interest groups takes place within the Technical Coordination Committee at the Provincial level. The committee includes the provincial governor, other ministries representatives and the farmer organisations represented by the Chamber of Agriculture (CdAg)

Constitutional accountability: ORMVAM is a public rural development organisation with financial autonomy under supervision of the Ministry of Agriculture and the Ministry of Finance. It has a Board of Governors (Conseil d'Administration) which includes representatives of national farmers organisations and other sector representatives.

Rijnland Waterboard, The Netherlands

Operational accountability: Executive management is responsible for the implementation of service delivery and reports directly to the management board. Charges are determined and approved by the Assembly, which consists of representatives from various stakeholders groups. Landowners are responsible for maintenance of collector drains adjacent to their land. Failing to do this, the Waterboard will carry out the necessary works and charge the expenses to the landholder.

Strategic accountability: The waterboard has a General Assembly whose members are beneficiaries of waterboard services. An Executive Board elected from members of the Assembly overviews the action of the Executive Management.

Constitutional accountability: The Water Authorities Act of 1991 provides the regulatory framework for the waterboards. The provincial governments are the competent authorities to establish/abolish the waterboards, determine their tasks and area of jurisdiction and form of members election and representation.

Chapter 5

Relation between Level of Service, Flow Control and Management

Everybody reasons about hydraulics, but there are few people who understand it ... For lack of principles, one adopts projects of which the cost is only too real but of which the success is ephemeral; one carries out projects for which the goal is not attained; one charges the state, the provinces, and the communities with considerable costs, without fruit, often with loss; or at least there is no proportion between the cost and the advantages which result therefrom .

P.L.G. Du Buat, 1786

In previous chapters, we discussed the concept of strategic planning and the concept of level of service including its formulation process. In this chapter, we will discuss the elements that enable the implementation of the agreed level of service -the hydraulic control infrastructure and the associated management arrangements- The main aim of this chapter is to describe the range of flow control technology that can be used in the provision of irrigation and drainage services and to describe its adequacy in relation to various types of level of service. The operation and maintenance of irrigation schemes must be based on the provision of an agreed level of service at an agreed level of cost. This concept applies to both government owned and managed systems and decentralised or privatised systems. The level of service must be decided by involving farmers through a comprehensive consultative process. During this process, a clear link between the level of service and cost of service provision must be established. This ultimately results in service agreements describing the transactions involved in delivering the services and the accountability mechanism related to implementation of the transactions. A given level of service can be provided by different hardware (infrastructure) and software (management inputs) combinations. The cost of service is determined by a combination of management, operation, maintenance and investment cost required to meet the specified service standard for water delivery and drainage.

5.1 IRRIGATION AND DRAINAGE LEVELS OF SERVICE

The ability to control the supply of water to individual farmers at the farm gate and dispose of excess water from rainfall or irrigation largely determines the success or failure of an irrigation system. In smallholder schemes, where the number of farmers is large, the water supply function is often performed through a Water User Association (WUA) at tertiary unit level. Because of the complexities involved both in the provision of water supply services and drainage services a clear set of rules and regulations for acquisition, conveyance, delivery, distribution and removal of water is needed. Moreover, an accountability mechanism needs to be in place to ensure that these rules are followed. However, these rules are made in different forums. In case of irrigation services, the irrigation and drainage agency develops a set of rules related to water allocation and delivery to the users or user groups. The user groups develop their own set of rules for distribution of water among their members. In the case of drainage services, the organisation may be required to develop differential levels of service for different customers such as those in urban, industrial and rural areas.

Hydraulic infrastructure and management inputs are the two key elements that the organisation relies upon to meet its level of service obligations. The combined use and share of each of these two resources in the delivery of service determines the ability to meet the operational objectives of the system. The type of hydraulic infrastructure in itself has inherent physical limitations in the way it can be operated. For example, manually upstream controlled systems are not well adapted for rapid response to changes in flow demand; however well developed operational procedures and highly skilled operators can offset some of these constraints and reduce the response time required to accommodate flow changes in the system.

Involving the features of the hydraulic infrastructure and the levels of management inputs must form an integral part in the development of the level of service to be provided by the organisation. The features of the hydraulic infrastructure can also evolve over time in response to the need for changes in the type and level of service required. Likewise, the skill level of the organisation's staff can be raised through appropriate training programs designed to support the implementation of management objectives related to the provision of services.

5.1.1   Irrigation Levels of Service

The water delivery concept refers to the way in which the flow delivery schedule is determined. The concept recognises two levels of definition: (a) *strategic level* and (b) *operational level*. At the strategic level a differentiation must be made according to the degree of freedom that users can exercise in withdrawing water from the supply system. In this context, three main categories of system operation can be identified:

+ *on-demand:* Users can extract water from the system at any time without prior request;

+ *on-request:* Users must order water in advance of the intended time of withdrawal;

+ *imposed:* Users can only obtain water at the time dictated by the agency.

There are three main variables that characterise the flexibility of the delivery at the operational level: the rate of flow, the duration of supply and the frequency of supply. A number of possible operational scenarios can be obtained from the combination of strategic and operational categories as summarised in Table 5.1. Operational variables can be further constrained according to whether the operational variables are constant, pre-determined (not modifiable) or variable (modifiable) over the irrigation season.

Flexibility, adequacy and reliability are the main qualities that can be used to measure and judge a given level of service. In this discussion, the level of service is equated with the degree of flexibility in water delivery and the degree of freedom that users can exercise in on the choice of the main delivery parameters: discharge, duration and frequency.

The highest level of flexibility is achieved with an *on-demand* system. Users can obtain water directly from the supply point at any time without having to inform the agency. Because of the limited capacity of the canal system, sometimes restrictions are imposed on the flow rate by limiting the maximum discharge that can be diverted, the duration of the delivery or the time at which the withdrawal can be made. The imposition of these types of constraints can be considered to reduce the level of service.

On-request service implies that users must order water from the irrigation authority to obtain water. The authority will evaluate the request and decide on the rate, time and duration of delivery, which can differ from that requested. Aspects included in the making this evaluation can be water availability, canal capacity, priority rights, consumed water

Table 5.1 Classification of levels of irrigation service (Hofwegen and Malano, 1997)

Class	Level of service	Operational variants
Ia	On-demand - unrestricted	users take water when needed in any rate and duration
Ib	On demand - one variable restricted	users take water when needed with limited rate or limited duration or with limited frequency
IIa	On-request - unrestricted	users request supply in terms of rate, duration and time
IIb	On-request - one variable fixed	users request supply in terms of rate and duration or rate and time or duration and time.
IIc	On-request - two variables fixed	users request supply in terms of rate or time or duration
IIId	Imposed – all variables predetermined	agency supplies predetermined rate at predetermined time for predetermined duration

rights, overlapping requests, travel time from the resource and outstanding debts. This process of evaluation and the travel time of water from the source normally requires a notice period for the authority to organise the delivery, which varies according to each particular scheme according to availability of water and hydraulic capacity of the conveyance network. Depending on the degree flexibility that users have to order water, they may be able to specify the discharge, time and duration of the delivery or limitations may be imposed on any of these variables. Because of the notice period involved and the uncertainty that the requested amount will supplied, on-request is considered a lower level of service than on-demand.

The lowest level of service is obtained from *imposed* deliveries. Under this type of service, the authority decides on the allocations and schedules based on certain criteria, which may involve some form of user consultation. Rate, duration and time are predetermined by the agency in the form of an operation schedule for a complete season and usually remain unchanged.

The highest level of service is not automatically the most appropriate or desired level of service either by the farmers or by the authority. Flexibility can also be increased by developing more infrastructure on the farm. Instead of having a costly but flexible delivery service from the authority a system of on-farm storage or conjunctive use of groundwater may provide the same desired level of flexibility. In such cases, farmers might prefer a cheaper but lower level of service from the irrigation agency. The level of service provided by the agency may in some cases not be the highest possible due to restrictions imposed by government policy. If water is scarce and allocation and distribution need to be controlled to supply water to all users then a lower level of service may have to be provided.

5.1.2   Drainage Levels of Service

The provision of drainage services encompasses a wide range of excess water sources and service beneficiaries. Drainage service is concerned with the removal of excess water due to surface runoff and the control of surface and groundwater levels from agricultural, nature and urban areas.

Surface Runoff

In most cases, the provision of drainage service is dominated by the removal of excess water from rainfall. In this regard, the level of service is directly related to the desire and ability to evacuate the excess water in the shortest time possible. Then the service criterion is usually based on the removal of a standard storm within a certain time that is converted into a design discharge per unit area usually termed *drainage coefficient* or *drainage module*. For example, the service criteria for drainage of paddy areas in Indonesia is the removal of the 3-day rainfall event with 5 year return period of 3 days expressed in litres per second per hectare [l/s.ha] (Ministry of Public Works, Indonesia, 1986). The level of service will be higher for a higher return period and a faster removal of excess water.

Drainage system capacity is the most important restriction on the level of drainage service. In sloping areas flooding risk at the downstream end of the drainage system and at the lowest spots will be higher as the capacity will already be taken up by runoff from the upstream area, if the canal capacity is the only limiting factor. To regulate this inequity, the service condition could provide a restricted capacity of the drainage outlets from the contributing fields or sub-drainage areas. Such a restriction would impact on the drainage capacity of the upper area served by the outlet, but will ensure a more equitable service throughout the entire area. The outlet capacity can be related to size of the drained area and the crop value or other assets built in the area. This type of arrangement provides a direct basis for establishing a relation between the level of drainage service and drainage charges - the greater the capacity provided by the outlet, the greater the capacity of the drainage removal system ensuring a higher level of service and higher the associated cost.

The storm run-off from urban areas within an irrigation scheme is usually higher due to the roofed and paved surfaces. Collector drains that are constructed for rural purposes are often unable to handle these peak flows. The requirement may either be met by increasing drain capacity or by constructing retarding or holding basins in the urban area to reduce the peak flows.

In flat tidal areas the drainage capacity is also determined by the capacity of the outlet: gravity sluices or pumping stations. This capacity has a direct relation with the storage capacity in the drainage area itself. An optimisation has to be made of the trade off between the storage area and the sluice or pump capacity. A higher level of service will mean a combination of a higher sluice or pump capacity and larger storage capacity.

Groundwater level control

Groundwater level control may be necessary to:

♦ prevent waterlogging in the root zone

♦ prevent saline or brackish groundwater water from entering the root zone

♦ allow crops to benefit from capillary water

♦ prevent oxidation of peat and acid sulphate soils

The optimum depth of the groundwater table is different for each situation depending on topography, crops and soils; and as such it would be most likely different for each field. The service criteria related to these control objectives are expressed according to the specific drainage purpose. Probability of exceedance of target groundwater levels may be used in situations where the water level equilibrium must be maintained to prevent soil oxidation processes. As the farmers manage most of the field drainage systems, the service criteria are translated in target water levels in the receiving collector drains or waterways with margins of variation at a certain probability. Where drainage service must be provided to control waterlogging and salinisation from irrigation, the depth of water in the collector drains and tolerance in fluctuations are used to define the drainage service.

The level of drainage service is higher if it can accommodate particular (ground-) water table requirements on a more detailed scale. This will require a more detailed control and management of infrastructure which will result in a higher cost of service. All these service level classes have their specific requirements for infrastructure (especially flow control and discharge measurement systems) and facilities and consequently on the investment and management cost involved. For all these categories special provisions must be made for the flow control systems.

5.2   FLOW CONTROL CONCEPTS

Flow control systems form the core of the irrigation and drainage infrastructure. The purpose of the flow control system is to regulate water flows and water levels in the canal network to meet the delivery specifications of the agreed level of service. These specifications are expressed in criteria and standards to be met regarding flexibility, reliability, equity and adequacy of delivery as described in Chapter 4.

There are several canal control concepts that are related to the ability and adequacy of the water control system to meet a specified level of service in irrigation and drainage. These include (a) the method of control, (b) the control logic, and (c) the field configuration of the system. A discussion on these concepts is provided in the following sections. It is important to note that the discussion focuses on those aspects related to the ability of the flow control system to meet specific level of service requirements. The reader may find a more detailed discussion of these concepts in various other specialised publications (Ankum, 1991, Malaterre, 1995, and Buyalski *et al*, 1991).

5.2.1   Flow Control Methods

The discharge and water level in open channel systems is subject to variations due to changes introduced during the operation of the system to meet water delivery requirements. The aim of the channel operator is to meet operational objectives designed to satisfy the supply requirements at various points in the system by achieving a certain steady state condition of the system, also called the "desired state" of the system. Several flow control methods can be used to achieve this objective including (a) Fixed division control, (b) upstream control, (c) downstream control and (d) volume control.

a. FIXED OVERFLOW SYSTEM   b.SUBMERGED UNDERSHOT SYSTEM

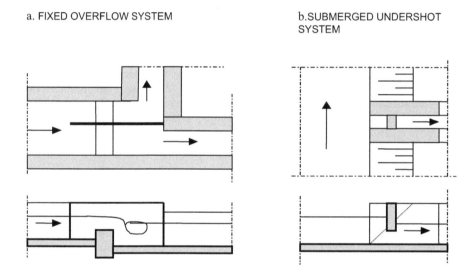

Figure 5.1 Fixed overflow and submerged undershot fixed division systems

Fixed Division Systems

A distinction in flow control has to be made between systems that achieve water division by using fixed division structures and those that have gated offtakes along the canal (Murray-Rust and Snellen, 1993). Fixed division systems are those where water can only be managed at the head of the canal; division between subsidiary canals or offtakes along the canal is achieved through fixed division structures that have no gates. Gated division systems are fitted with gates at each offtake along the canal to allow water to be regulated at every bifurcation in the system. Generally water level regulators are used to maintain constant water levels so that offtake flows can be properly controlled. Two different types of fixed division systems are available (fig. 5.1):

- ◆ *fixed overflow weirs:* The width of each weir section is proportional to the supplied water rights based on a percentage share of the available water. This control system is also referred to as *proportional* control. These systems provide a constant distribution of water shares over a wide range of discharge;

♦ *submerged orifices:* These are designed to deliver a relatively constant discharge over a limited range of operating heads. The proportional distribution in these systems is conditional to the water level in the parent canal being maintained within a certain margin of the full supply level. Typical examples of this type of systems are the Mogha structures used in Pakistan. In these canals the water level is not controlled by cross regulators but only determined by the normal depth of flow in the canals. During operation, the canal runs at or near full discharge level.

These systems are particularly effective in meeting equity objectives based on a percentage share of available water on a per unit area basis, per household or per person basis, as long as the water shares remain constant. This means that the system is rigid and can not easily respond to agricultural changes such as expansion of area or change in the number or size of streams. As these systems are supply based, there is little control to permit discharges to be adequately managed to meet variable demands. Individual farmers or water users groups adjust their demands to water availability through careful selection of the cropping pattern. Fixed division systems are generally predictable; however, the variability of flows is highly dependent on the flow variability at the source.

Upstream Control

Upstream control is by far the most commonly used type of flow control around the world. With upstream control, the water level in the canal is controlled upstream of the regulator. Figure 5.2 illustrates this form of control. Changing the discharge requires changing the setting of the gates upstream of the location concerned. It provides agencies with a relatively high level of control of deliveries such as in the case of imposed or scheduled delivery. Upstream control systems are supply oriented, have limited flexibility and require agency management.

Two types of arrangement can be identified based on the type of cross regulators which also determine the degree to which the upstream canal water level can be controlled:

♦ *fixed cross regulation:* Overshot weirs typically result in stable head-discharge conditions. Although this type of cross regulators is fixed, variable water levels can be achieved by using removable stoplogs;

♦ *gated cross-regulation:* Gates in the canals itself can be used to manage water levels irrespective of the discharge. The vast majority of these systems are manually controlled but more modern systems utilise hydraulic or electrical automation systems. Offtake gates play a critical role in delivering water to meet level of service specifications. The mode of delivery can determine the selection of the appropriate offtake gate.

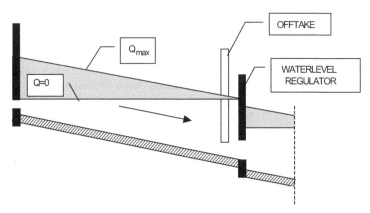

Figure 5.2   Upstream Control

For instance, if water is to be supplied by fixed discharge and variable duration, fixed undershot gates or on-off gates can be appropriate as they provide a constant discharge over a range of water levels upstream of the gate. If the level of service involves variable discharge and set duration, then movable gates in combination with measuring devices are required. For both options, it is important that the water level in the parent canal is maintained within a certain range of fluctuations. The combination of offtake and water level regulator will determine the range of fluctuations of water levels in the parent canal. This range of tolerance for water level variations must be specified in the service agreement. As the variations permitted by the service agreement become smaller, more amount of management inputs are required in the form of more frequent monitoring and operational adjustments.

Systems with gated cross regulators are more flexible than those with fixed regulation facilities. They are essential to accommodate water operation plans that are more responsive to demand. With these systems, it is possible to manage adequacy more closely as they allow a greater control of supply to meet short-term changes in demand. They also have the potential to be very reliable both in terms of predictability and variability. However, if poorly managed, there is a potential for very high levels of inequity and unreliability.

Downstream Control

In downstream control the water level is controlled on the downstream side of the regulator as illustrated in Figure 5.3. Downstream control systems are designed to permit instant response to changes in demand. Most downstream control systems use balanced hydraulic gates that open or close in direct response to changes in the water level downstream of the structure. Thus the system becomes self-regulating. Downstream control significantly reduces response times and the associated operational losses. Also the management input can be reduced. However, the application of downstream control requires a higher investment cost due to more expensive gates and a requirement for horizontal canal banks. This requirement limits the use of this type of control where gradients are mild (0.2 – 0.3 m/km). Downstream control is not appropriate when the total allocated deliveries exceed the water availability. In such situations inequities arise as the system responds to downstream users' demand leaving upstream areas undersupplied.

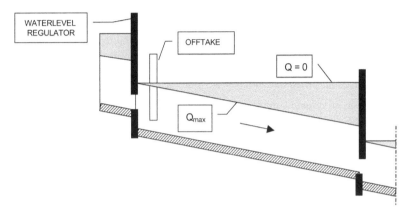

Figure 5.3 Downstream control

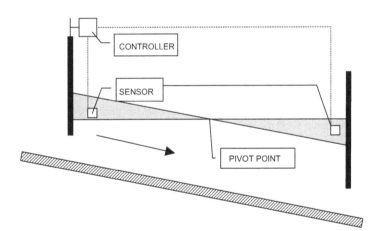

Figure 5.4 Volume control

Volume Control

The volume control method regulates the volume of water in each canal pool by providing an instantaneous response to any changes in the hydraulic state.   This involves a continuous monitoring of the water level throughout the canal and the simultaneous operation of all control structures to maintain a nearly constant volume of water in each canal pool. With this type of operation, the water surface between control structures rotates around a pivot point located approximately midway between the control structures as illustrated in figure 5.4.

This method is used to meet operational requirements for different users of water in a flexible manner as it can provide immediate action to respond to changing demand conditions. Implementing this type of control requires automation of the control structures coupled with centralised monitoring and control. This involves a network of remote sensors and transmitters, which collect information and relay instructions back to the control structures and pumps throughout the system. The control logic is exercised through an appropriate computer controlled algorithm. The actual operation of the appurtenant structures is carried out with electric motorised controls.

The main advantage of volume control over downstream control is that the embankments can be lower and the system can be applied in steeper canals resulting in lower construction costs.

5.2.2   Control Logic

The concepts of control systems engineering applied to mechanical engineering systems can also be applied to the control of irrigation and drainage canal systems. The nature of open channel flow is such that flow transients develop every time the setting of the control structures is changed to meet operational objectives. The control logic employed by the system determines the manner in which the system achieves the desired state selected to meet level of service objectives.

The main objective of the control system is the control of the *flow process* based on the specific operational objectives such as water level or discharge, which are subject to external *perturbations*. These result from changes in the discharge and depth due to extractions or augmentation of flow in the canal and changes in settings of the control devices. The control system consists of the hardware and software components needed to modify the flow parameters. These include the *sensor*, the *comparator*, the *controller,* the *actuator* and the *regulator*. All these functions can be performed manually or by mechanically automated equipment. For instance, the sensor task can be carried out by a staff gauge or by an electronic water depth sensor; the actuator function can be performed by a gate operator or by a motorised gate lifting mechanism. The regulator is the structure that governs the controlled process such as a proportional divisor, a slide gate, a overshot weir device or a pump.

There are two main forms of control logic used in open channel systems: (a) closed-loop or "feedback control" and (b) open-loop or "feed-forward control." Figure 5.5 shows an illustration of these two control criteria. Under closed-loop control the process output variable (flow depth or discharge) which varies as a result of perturbations in the system is compared with a pre-set target value. The controller processes deviations from the target point and issues correcting instructions to the actuator to carry out the adjustment of the regulators. This is a continuous process of adjustment of the system structures to drive the system towards the desired state. While these functions can in theory be carried out either manually or mechanically automated, the ability of field operators to perform these adjustments is often limited by their physical capacity and the degree of stability of the hydraulic system. Feedback control can be implemented in conjunction with local control or central control as detailed in section 5.2.3.

In open-loop systems, the desired state of the system is determined by criteria based on the specific level of service to be delivered. The system controller and regulators operate in the same manner as in a closed-loop system except that now the output process variables is not adjusted in response to deviations from the target value due to perturbations in the system. This is often the case in manually controlled systems where the field operation staff sets the   control structures according to instructions received from the central operation office and no further monitoring and adjustments are carried out. However, in this case there are no provisions to carry out corrective changes in the operation if the outcome of the operation is significantly different from what was intended.

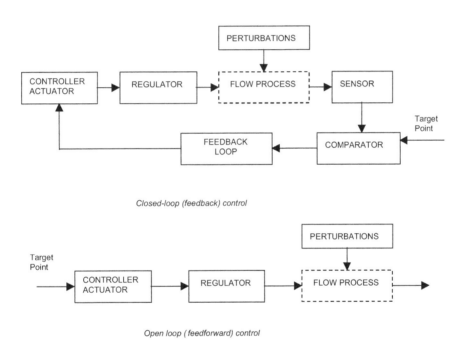

Figure 5.5 Schematic illustration of closed-loop and open-loop control systems

5.2.3   System Configuration

The field implementation of the control system involves a number of links between controllers, actuators and control variables. These elements can be arranged in various ways considering their spatial distribution (Buyalski *et al*, 1991): (a) local manual control, (b) local automatic control, (c) supervisory control, and (d) combined methods.

Local manual control is the most commonly used method of canal control in irrigation. A human operator who may be assisted by on-site mechanical devices such as motorised gate lifting mechanisms carries out the task of the actuator. The field operator usually acts on instructions issued by the operation control centre to adjust the setting of the canal structures in order to meet the service requirement. This type of control method is complex and requires good communications and detailed operational procedures. The success of the operation depends to a large extend on these factors although in large canals the operation may become exceedingly complicated limiting the quality of service provided. A closed-loop system can be implemented if the field operator monitors and adjusts the setting of the structures at certain intervals. Manual control imposes some limitations on the flexibility of timing and changes in supply discharge. Moreover, the tolerance set for the level of service specifications such as fluctuations in supply must consider the limitations of this type of operation.

Local automatic control relies on on-site control equipment to carry out the operation. This type of control commonly relies on closed-loop feed back systems to provide automatic control of flow depths. A number of commercially available devices have been developed for local automatic control both for upstream and downstream controlled canals, e.g. AMIL gates, Little Man, ELFLO, AVIS/AVIO, etc. (Malaterre, 1995). Under these types of control structures, human intervention is not required for flow adjustments although periodic monitoring and maintenance inspections are necessary. If properly installed and maintained, locally automated regulators enable the irrigation organisation to deliver a higher level of service and to meet more stringent level of service specifications.

The operation of the canal system by *central supervisory control* is carried out remotely from a control centre, which centralises all the canal system information in one location. This type of control operation is usually termed Supervisory Control and Data Acquisition (SCADA). It involves monitoring the state of flow at multiple points in the systems (thus supervisory) and controlling actions needed to regulate the multiple control structures in the system. All monitoring and control actions are carried out in real time with either a human operator or a computer driven algorithm determining the structure settings to meet the desired target state of the system. The operator usually relies on appropriate computer algorithms to determine the control actions required to correct the flow depth or discharge on the basis of the monitoring information. Because the system is centrally controlled, the system operator can reduce the response time of the system by acting simultaneously on multiple control structures and pumps throughout the system.

Systems that use the volume control method of flow control must rely on fully integrated computerised algorithms to operate structures and pumps throughout the system such as in the Canal de Provence, France (Rogier, 1987)**.** On the other hand, conventional upstream control systems fitted with SCADA may rely on human control assisted by appropriate computer-driven decision support systems.

Often the control of a canal system relies on a *combination of control methods*. Two of more of the control methods described above can co-exist in a canal system. This situation often arises in upgraded canal systems where the main system is centrally controlled and lower canals are locally controlled either manual or automatically.

5.3   WATER CONTROL IN DRAINAGE SYSTEMS

The operational objectives required for the provision of drainage service contrast significantly with those of irrigation. As discussed in Section 5.1.2, drainage services can be aimed at the removal of water from excess surface runoff or at controlling the groundwater level, which may rise due to excess rainfall or excess irrigation. In both cases, the function of the main drainage system is to convey the drainage water to the outlet of the system, and under certain conditions, control the water level for crop water supply or for prevention of soil oxidation processes.

Upstream control is commonly used for flow control in drainage canals that require a controlled level for navigation or groundwater management. Control of water level in drainage canals is attained by controlling the water level upstream of the control structure or at the downstream end of the pool.

Downstream control has fewer applications in drainage canals. It can be used when a reduction or attenuation of the peak discharge from the drainage system is required. Downstream control will force a positive storage of water in the canal system or force the canal to spill into another holding canal or storage. A higher water level in the canal may also limit the drainage outflow from the drainage service areas. The selected level of drainage service must be specified in terms of the drainage capacity or drainage coefficient.

5.4   LEVEL OF SERVICE AND FLOW CONTROL SYSTEMS

Some flow control systems are more appropriate than others to meet a specific level of service. The adequacy of a particular flow control system to deliver a specific level of service is determined by the relation between the type of system operation that is required and the hydraulic behaviour of the flow under that particular flow control system. In general, as illustrated in table 5.2, hydraulic control methods that respond more flexibly to flow variations caused by users withdrawals are better suited to provide a more flexible level of service. For instance, automatic downstream control systems are well suited to respond quickly to changes in the flow rate. This makes it possible for users to withdraw water without having to notify the operating agency as flows can readily be adjusted automatically provided there is no constraint on water availability. On the other hand, a manually controlled system would take longer to respond to these changes and the fluctuations in the system would also be greater. The same level of service, however, can be achieved with different technologies. It is also possible to achieve the same flexibility and responsiveness with a centrally control supervisory system. The choice of the technology ultimately becomes an exercise in sound engineering and management judgement for a given set of physical, financial and socio-economic conditions.

On-request service can be provided with a range of flow control systems albeit with different degree of adequacy depending on the degree of flexibility required. Upstream manually controlled systems have more difficulty in delivering on-request service with no constraints on rate of flow, frequency or duration given that a large amount of canal regulation is involved. Often this type of service arrangement compels operators to run the system with greater than required discharge to reduce the risk of not being able to accommodate water orders within the notice period resulting in greater operational losses at the end of the system. Operators are usually better able to cope with on-request service if one or two of the delivery parameters are constrained.

The delivery of a specific level of service has specific staffing requirements, which must be given appropriate consideration. In principle, each design has intrinsic staffing requirements in terms of number and level of skills. The average staffing requirements and associated cost for various types of level of service are summarised in table 5.2. It must be recognised however, that while the classification presented in this table attaches discrete degrees of adequacy and staff requirements to each type of delivery service, in reality adequacy levels vary in a rather continuous fashion. Staff skills and competence are the main factors that determine the staffing level and the staff ability to operate the system.

5.5 RELATION BETWEEN FLOW CONTROL AND MANAGEMENT INPUTS

A certain amount of substitution may exist between staffing requirements and hydraulic control depending on the type of service that must be provided with a given hydraulic infrastructure. In theory, there is some flexibility in selecting a particular type of flow control system to deliver a specified level of service as several types of flow control could be adequate to deliver that level of service. A trade-off between the type of flow control technology and management inputs provides this flexibility. It must be stressed, however, that this ability to substitute hardware for management inputs can only take place within certain limits. For a particular type of less flexible flow control such as proportional control-fixed division systems, it is not possible at all to deliver some specified levels of service including on-demand or on-request services.

Designs that include adjustable structures require operational inputs to carry out the adjustments in the setting of these structures. Fixed systems can only be operated at the control locations at the head of each major canal thus requiring minimal operation. Gated division systems require greater operational inputs because every offtake structure and its associated water level regulator must be adjusted frequently.

These inputs increase in direct proportion to the number of adjustable regulators. Adjustments however, can be made manually or automatically, locally or centralised via remote control.

Achieving flexible delivery with upstream control systems requires additional staff inputs and good communication arrangements to cope with the more intensive canal regulation requirement to change the flow conditions in the canals from one flow state to a new stable flow state. Remote centralised control is particularly suited to overcome these constraints and to increase the flexibility of delivery. Downstream control systems can respond almost instantaneously to varying conditions in the flow caused by water withdrawals by users.

Manual control can be as effective as automatic control in delivering a set level of service. This represents the more clear case of trade off between hardware and staffing level provided that there are no constraints to increased staffing level and communication. Automatic control can reduce the staffing requirements in terms of number but is more demanding in the level of expertise and skills required of the staff.

Table 5.2. Adequacy and staffing requirements for various flow control systems and levels of service.

Level of Service		Operation Parameters			Flow Control System.					Human Resources.	
Water delivery	Class	Rate	Dur	Freq	Prop	UC Man.	UC Aut. Loc	UC aut. cent	DC aut	HR No	HR skill
ON-DEMAND											
Unrestricted	Ia	V	V	V	N.A	--	--	o	++	++	o
Limited rate	Ib$_r$	C	V	V	N.A	--	--	o	++	++	o
Limited duration	Ib$_d$	V	C	V	N.A	--	--	o	++	+	o
Limited frequency	Ib$_f$	V	V	C	N.A	--	--	+	++	+	o
ON-REQUEST											
Unrestricted	IIa	V	V	V	N.A	--	-	+	++	--	-
Fixed rate	IIb$_r$	C	V	V	N.A	-	o	++	++	o	o
Fixed duration	IIb$_d$	V	C	V	+	-	o	++	++	-	o
Fixed frequency	IIb$_f$	V	V	C	+	-	o	++	++	o	o
Fixed rate+duration	IIc$_{rd}$	C	C	V	N.A	O	+	++	++	o	o
Fixed rate+frequency	IIc$_{fc}$	C	V	C	N.A	O	+	++	++	o	+
Fixed duration +frequency	IIc$_{df}$	V	C	C	+	O	+	++	++	o	o
IMPOSED											
Fixed rate+duration	IIIc$_{rd}$	C	C	V	N.A	O	+	++	++	+	o
Fixed rate+frequency	IIIc$_{rf}$	C	V	C	N.A	O	+	++	++	+	o
Fixed duration +frequency	IIIc$_{df}$	V	C	C	+	O	+	++	++	+	o
All fixed	IIId	C	C	C	N.A	+	+	++	++	+	+

prop = proportional,   UC = upstream control,   DC = downstream control,   HR = human resources;
aut = automatic,   loc = local;   cent = central;   No = number;
C = fixed;   V = variable;   N.A. = not applicable

-- very unfavorable   - unfavorable   o neutral   + favorable   ++ very favorable

5.6   THE LINK BETWEEN LEVEL AND COST OF SERVICE

In previous sections, the possibility of providing different levels of service with various types of flow control technology was discussed. As a result we can conclude that one level of service can be provided with several types of flow control, and conversely, one type of flow control can be used to provide different levels of service. However, the provision of a higher level of service with a given type of flow control requires additional staffing with greater skills and proficiency for planning and executing the system operations. In other words, to provide a given level of service, flow control and management can substitute each other within a certain limit.

Table 5.3 illustrates the relation between the relative management and infrastructure costs associated with different levels of service. As will be discussed in Chapter 6, the costs associated with the provision of a specific level of service are operation, asset maintenance, asset depreciation and return on investment. It is however unusual for government owned and operated irrigation schemes and even for privatised schemes to account for a rate of return on the irrigation and drainage investment.

Operation costs are primarily determined by the number and skill level of the staff required for the operation of the flow control infrastructure and associated plant and equipment.

There is an inverse relation between the level of automation of flow control and the number of staff required to operate the system. However, a higher level of skills is required to operate a system that relies on a greater level of automation or relies on conventional upstream control technology to provide a flexible service.

Conventional upstream flow control technology is used worldwide by many irrigation agencies to provide a wide range of levels of service. In Australia, several irrigation districts in the Murray-Darling Basin provide on-request unrestricted water supply service using the same basic and conventional upstream flow control technology. This level of service is achieved by way of highly skilled operation staff and well-developed operational procedures. In the Triffa scheme in Morocco, a combination of upstream control with local automation is used to provide a lower level of service but at higher efficiency (Box 5.1). These are examples of the possible trade-offs that can exist between flow control technology and management.

This trade off between flow control technology and management also has implications on the operational efficiency of the system. In theory, it is possible to supply water on-demand with a manually operated upstream controlled system by continuously operating the system at maximum capacity regardless of the actual demand. In practice however, such an operation will result in significant wastage of water when demand is low, for instance after a rainfall.

The number of flow control structures and their individual cost primarily determine maintenance costs. Mechanically complex structures generally have higher maintenance costs and are more sensitive to maintenance standard. This is significant for fully automated systems that rely on computerised control and a large number of sensors and controllers, which required periodic routine maintenance and can be easily vandalised. Moreover, cost of spare parts is greatly affected by whether specialised services are available domestically or they must be procured overseas. This often creates additional logistical difficulties in some countries due to foreign currency restrictions.

Depreciation is usually the largest cost item and is intended to reflect the rate of consumption of the infrastructure. The initial cost and the service life of the infrastructure will determine the magnitude of depreciation.

Flow control systems that involve a higher level of automation require a greater initial investment. This does not necessarily imply that the depreciation cost will be greater provided the service life is not shortened by inadequate maintenance.

The trade off between infrastructure investment and management to meet a certain level of service as discussed here is based on the assumption that the level of service has been defined as part of a comprehensive strategic plan for the agency. As such, it is founded on the principle that a training program will be in place to support the strategies designed to achieve the strategic goals and guarantee that the staff will acquire the appropriate skills to operate and maintain the type of infrastructure chosen for flow control.

Table 5.3. Level of Service (LOS) and Management Inputs for Different Flow Control Systems.

LOS	Proportional Staff No	S	Infrastr. I	M	W Eff	Upstream Control Manual Staff No	S	Infrast. I	M	W Eff	Upstream Control Automatic-local Staff No	S	Infrast. I	M	W Eff	Upstream Control Automatic-central Staff N	S	Infrastr. I	M	W Eff	Downstream Control Automatic-local Staff No	S	Infrastr. I	M	W Eff
Ia	na	na	na	na	na	--	--	o	-	-	-	--	-	--	-	o	--	--	--	o	++	o	--	--	+
Ib$_r$	na	na	na	na	na	--	-	o	-	-	-	--	-	--	-	o	--	--	--	o	++	o	-	--	+
Ib$_d$	na	na	na	na	na	--	-	o	-	-	-	-	-	--	-	o	--	--	--	o	++	o	-	--	+
Ib$_f$	na	na	na	na	na	--	-	o	-	-	-	-	-	--	-	o	--	--	--	o	++	o	-	--	+
IIa	na	na	na	na	na	-	-	o	-	o	o	o	o	--	o	o	--	--	--	+	+	-	-	--	++
IIb$_r$	na	na	na	na	na	-	o	+	-	o	o	o	o	-	+	+	--	-	--	++	+	-	-	--	++
IIb$_d$	++	+	++	++	-	-	-	o	-	-	o	-	-	-	+	+	--	-	--	++	+	-	-	--	++
IIb$_f$	++	+	++	++	-	-	-	o	-	-	o	-	-	-	+	+	--	-	--	++	+	-	-	--	++
IIc$_{rd}$	na	na	na	na	na	o	o	+	o	o	+	-	o	o	o	+	--	-	--	+	+	-	o	-	+
IIc$_{rf}$	na	na	na	na	na	o	+	+	o	o	+	-	o	-	o	+	-	-	--	+	+	-	o	-	+
IIc$_{df}$	++	+	++	++	--	o	o	o	-	-	o	-	-	-	o	+	-	--	--	+	+	-	-	-	+
IIIc$_{rd}$	na	na	na	na	Na	+	+	+	o	-	+	o	o	-	o	+	-	-	--	o	+	-	o	-	o
IIIc$_{rf}$	na	na	na	na	Na	+	+	+	o	-	+	o	o	-	o	+	-	-	--	o	+	-	o	-	o
IIIc$_{df}$	++	++	++	++	--	+	++	o	-	o	o	o	-	-	--	+	-	-	--	o	+	-	o	-	o

-- very unfavorable - unfavorable o neutral + favorable ++ very favorable ; na = not applicable

N = number; S=Skill; I= Investment; M = maintenance; W = water; Eff = operational efficiency; na = not applicable.

Box 5.1 Comparison and analysis of the level of service, water control technology and management of two irrigation schemes: The Triffa irrigation scheme in Morocco and the Central Goulburn irrigation area, in Australia.

TRIFFA IRRIGATION SCHEME	CENTRAL GOULBURN-IRRIGATION AREA

Description

The Triffa scheme (36,160 ha) is located on the Right Bank of the Moulouya river and is one of the four schemes (total irrigated area 71,000 ha) in the Moulouya basin in the North Eastern Morocco (van Hofwegen *et.al.* 1996b). The climate is semi-arid Mediterranean with a low and irregular average annual rainfall of 300 mm varying between 150 mm and 450 mm. Rainfall is concentrated in the months of December, January and April. The area is supplied from the Moulouya River with an average annual runoff of 800 MCM regulated by the Mohammed V and the Mechraa Homadi Barrage Periods of scarcity occur frequently. At Moulay Ali, downstream of the Mechraa Homadi Barrage, a pumping station was built for additional supply tail water with a capacity of 4 m³/s. It allows an additional average annual volume of 100 MCM to be pumped up some 95 meters from the Moulouya River into the Triffa Main Canal. The Triffa plain is further supplied with groundwater of good quality and sufficient quantity. It is used in times of scarcity for supplementary irrigation.

The Central Goulburn Irrigation Area is situated in the state of Victoria in the most southern part of the Murray Darling Basin (Goulburn-Murray Water, 1995a). The Goulburn River bounds the area to the East and North. The climate is semi-arid with an average rainfall of 400 mm varying between 250 mm and 900 mm. Rainfall is uniformly distributed over the year, a factor which maximises the need for irrigation during the summer months. The area is supplied from Eildon reservoir on the Goulburn River with a capacity of 3390 MCM. Reservoir releases are conveyed via the Goulburn River to Goulburn Weir where flows are diverted into two main feeder canals; the Stuart-Murray canal and the Cattanah canal. The development of the system commenced early in this century and was completed in 1955.

Hydraulic Infrastructure

Water is diverted through the Mechra Homadi barrage, which consists of a float water level regulator in combination with a battery of baffles. Regulation of intake discharges is carried out in steps of 500 l/s. The main canal also supplies several communities with a total of 500 l/s water for sanitation and drinking water. The main canal is concrete lined and has a length of 155 km. Water regulation in the main canal is carried out by a combination of control structures including duckbill weirs, mixed control Neyrpic gates, AVIO Downstream Control Gates and AMIL Upstream Control gates. The turnouts into the secondary canals consist of Neyrpic Baffle distributors. The lower level conveyance canals in the secondary units are elevated semi-circular concrete canals. The lower level turnouts (*prises*) are Neyrpic Baffles with the water level regulated by duckbill weirs. Water distribution between the farmers downstream of the prises is carried out through on-off gates in distribution boxes.

The hydraulic infrastructure of the Central Goulburn Area consists of a typical upstream control system which relies on a combination of gated cross regulators on the main canal and movable weirs and gated cross regulators in the lower rank canals. Water distribution to farmers takes place through a combination of gate and Dethridge wheel meters that enable the volumetric metering of deliveries. More recently, some key cross regulating structures have been motorised and controlled by a SCADA technology.

Institutional Arrangements

Management is the responsibility of the Regional Agricultural Development Bureau of Moulouya (ORMVAM – Office Regional de Mise en Valeur Agricole de la Moulouya) established in 1966 as a public authority with financial autonomy under the Ministry of Agriculture. Water charges are collected by ORMVAM and are largely used to offset the operational costs. Water proces are dictated by the national government leading to revenue shortfalls that are covered by subsidies from the Ministry of Finance. Investments on expansion or rehabilitation are financed from loans taken by the Government from international lending agencies.

Management is the responsibility of the Goulburn-Murray Rural Water Authority, a public corperation created in 1994 to succeed the former state water authority. Irrigation services are now provided on a full cost recovery basis including provisions for future replacement of existing assets on the basis of a 20-year rolling program. The asset renewal provision accounts for a significant proportion (44%) of the total price of water. The current water pricing accounts for all the actual costs incurred in providing the irrigation and drainage services. An asset management program provides the basis for the identification of infrastructure costs on an asset renewal basis to ensure the indefinite sustainability of the infrastructure.

Service Agreements

The service relation between the farmers and the ORMVAM is established through the water masters (*aiguadiers*). Farmers interests are represented by an elected Agriculture Council (*Chambre d'Agriculture*) acting as a member of the Governing Council of ORMVAM and the coordinating Provincial Technical Committee (CTP). While there is no formal contract governing the relation between agency and farmers, these can access water services only if they are officially registered landowners, tenant or representatives within the command area (*Depositeurs*) and if the land was purchased in accordance with the official regulations (code des investissements agricoles). Presently farmers obtain a statement ("*Deposition*") that allows them to obtain irrigation services by taking part in the irrigation cycles or "*Tour d'eau*". This constitutes an agreement that compels farmers to pay the water charges for a period of at least three years.

The relation between the Goulburn-Murray Rural Water Authority and farmers is governed by an agreement between the agency and the Water Services Committee that represents the customers (Goulburn-Murray Water, 1995b). The agreement does not create legal obligations on behalf of either party, but rather develops a partnership between the Authority and the customers. Legal obligations of the authority and farmers are specified by the Water Act, Environmental Protection Act and Occupational, Health and Safety Act which prevail regardless of the agreement between the Authority and Water Service Committee. The interests of customers are also represented on the Corporation's Board.

Level of Service

The basic principle of water distribution is that each land title owner (*TF: titre foncier*) receives a predetermined volume of water per irrigation cycle (*Tour d'eau*) usually in one but sometimes in two or more issues. This volume depends on the crop, the time of year and the reservoir level. Water is supplied at a fixed rate (20 or 30 l/s) through an ORMVAM managed turnout (*prise*) to a number of farmers. The farmers receive the full flow for the duration agreed with the *aiguadier*. ORMVAM decides on the implementation of the irrigation cycle, its duration and the volumes per hectare for the various crops depending on the availability of water. Farmers sign a note of acceptance of their share of water (*Feuille de Control*) which specifies the date, time, duration, discharge   and total volume delivered which will be used for assessing the water charges.

Farmers in the Triffa scheme are charged for irrigation services based on the volume of water delivered at their turnout according to the price set by the National Government. This price is used throughout Morocco and for the Triffa case, does not cover the full cost of system operation and maintenance. The average cost of water is about US$ 0.03 per m^3.   Additional charges are levied for lift irrigation (US$ 0.015 per m^3) and pressurised sprinkler irrigation (US$0.022 per m^3). At present no amortization is included in the price. (Van Hofwegen e.a., 1996)

The standard service that is provided by the Goulburn Murray Rural Water Authority consists of unconstrained on-request delivery where customers can select the discharge, starting and finish times of irrigation. Customers are required to provide four days notice for all orders with day 1 being the day after the order is placed. Although the agency allows for shorter notice periods, priority is given to orders placed with the required notice. A performance monitoring system is in place to ascertain the level of customer satisfaction and achievement of level of service goals. The service agreement between agency and users established a target of 86% of orders delivered on the day requested by users against which annual service performance is evaluated.

The price of water in the Central Goulburn Irrigation area is US$ 0.021 per m^3 (Central Goulburn, 1998). The water pricing structure has two components. Customers are charged for the fixed component of the water right entitlement and additional charges are calculated on the basis of the volume of sales water utilised. The volume of sales water allocation varies from year to year depending on weather conditions. The current price of water is based on full cost recovery including an annuity component for long term asset renewal.

Characteristic	Triffa Scheme-ORMVA de la Moulouya - Morocco	Central Goulburn Irrigation Area - Australia
1. Climate Annual rainfall	Mediterranean - Semi-Arid, 250 mm	Semi-arid 450 mm
2. Water source	Mohammed V Reservoir	Eildon Reservoir
3. Command Area Effective Area	36,000 ha 36,000 ha	173,050 ha 113,400 ha
4. Hydraulic Infrastructure -Main Canal: -Secondary Canals: -Tertiary Canals:	Lined trapezoidal Semi circular elevated conduits Lined/earthen canals	Earthen Earthen Earthen
5. Flow control - turnouts Water level regulation - tertiary canals - secondary canals - main canal	Neyrpic Modules a Masque Upstream control - on-off gates Upstream control - Fixed weirs Mixed upstream and downstream control - partly automatic	Gate + meter wheel Upstream control, movable weirs Upstream control, flat gates and weirs Upstream control, sluice gates and weirs
6. Irrigation Methods	Gravity - basin traditional (79%) Lift - basin traditional (20%) Pressurized (1%)	Gravity - borders (95.5%) Pressurized (4.5%)
7. Main Crops	Citrus: 42% Vegetables: 28% Wheat: 21% Others: 9%	Pasture : 96% Horticulture &vegetables 4%
8. Type of Organization	Public Corporation	Public Corporation
9. Level of Service	On request fixed rate, fixed frequency, variable duration	On request, unconstrained
10. Cost of Service (1996)	US$0.030 per m^3	US$ 0.021 per m^3
11. Cost Recovery	Partly cost recovery. Subsidies to cover deficits.	Full cost recovery including asset renewal annuity.
12. Delivery Performance	Target: delivery in accordance with agreed tour d'eau.	Target: 86% of orders delivered on day requested
13. Dispute Resolution	First instance: customer-agency reconciliation, otherwise regular judicial system.	First instance: customer-agency reconciliation, otherwise regular judicial system.

Chapter 6

Management of Irrigation and Drainage Infrastructure

The recurring cycle of construction-deterioration-rehabilitation(construction) usually precludes the timely and equitable distribution of irrigation water supplies. Besides resulting in stagnant growth in agricultural production, the construction phase adds to the national debt burden, which has now become a painful problem for many countries.

G. Skogerboe, 1990

Unlike industrial assets, irrigation and drainage infrastructure assets have very specific features such as dispersion and high cost in relation to the financial turnover of the managing organisation. Moreover, irrigation and drainage assets have specific hydraulic characteristics in relation to the function that they have to perform. Irrigation and drainage organisations have traditionally assigned little importance to the adequate management of the infrastructure. However, it is critical that as organisations are required to treat irrigation and drainage as a service business, they pay detailed attention to the cost of providing services. The cost of assets usually represents the largest cost associated with the provision of service. In order to provide this service in a sustainable and cost-effective way, a life-cycle approach must be taken to manage the organisation's assets. This involves the implementation of an asset management program that consists of several fully integrated functions including: asset creation planning, operation and maintenance, asset condition and performance monitoring and asset audit and renewal. The asset management program must be viewed as one of the organisation's plans within the overall strategic planning of the organisation. As such, it must be given proper and detailed attention must be paid to its development and implementation.

6.1   RATIONALE FOR AN ASSET MANAGEMENT PROGRAM

Irrigation and drainage authorities rely on their infrastructure to fulfil their level of service obligations. The irrigation and drainage infrastructure consists of a large number of individual assets including dams, canals, control structures, pumps, etc. However, unlike industrial production and commercial assets, they are usually dispersed over large areas. It is not uncommon that irrigation and drainage schemes extend over tens or even hundreds of kilometres. As explained in Chapter 5, a distinctive characteristic of irrigation and drainage assets is the close relation between the hydraulic configuration and management of the assets in question and the function that the asset is expected to perform in the provision of services. This intense reliance on the hydraulic infrastructure is usually accompanied by a relatively low financial turnover from irrigation and drainage charges.

This latter feature poses a greater challenge to irrigation and drainage managing organisation than to other urban water suppliers or industrial organisations in general.

Despite the importance of the asset base, most irrigation and drainage organisations place insufficient emphasis on its management. This stems from the fact that very few organisations have in the past been operated as a service business that requires accountability to customers, government and other stakeholders. Accurate information on the asset holdings except at an aggregated level is normally not available. With increased pressure on irrigation organisations to improve their level of self-sufficiency, it becomes imperative to be able to clearly quantify the cost of service provision. As pointed out earlier, the large asset base involved in the provision of irrigation and drainage service accounts for most of this cost.

Unlike other industrial and commercial activities, irrigation and drainage infrastructure consists predominantly of fixed assets, e.g. weirs, canals, control structures, drains, etc. The economic lifetime of these fixed assets is in the order of 20 – 50 years and for major assets like weirs and dams even longer, e.g. 100 years. The implications of this feature on the long-term maintenance and operation of the assets are very significant. The demands of irrigated agriculture change within such a timeframe due to changes in markets, cropping patterns and agricultural practices. The need for modernisation of agricultural practices is caused by an increasing pressure on profit margins due to globalisation of the markets and reduced government subsidies. This leads to greater investment in agricultural technology often accompanied by an increase in the size of the land holdings. Irrigation and drainage as service activities must be able to accommodate these changes by adapting the type of services provided often within the life of the existing infrastructure. To achieve this, the organisation must be able to conduct extensive forward planning of the changes required in the infrastructure. In doing so, it should consider the potential new operational requirements and the technological advances that can be adopted to improve both the service provided and reduce the overall cost of service provision.

The financial management associated with the provision of irrigation and drainage services varies between organisations due to a number of factors. These include the sources and availability of funds for maintenance and other functions, the age, condition and nature of the assets. Traditionally, financial management of most irrigation and drainage authorities is based on the budgetary allocation that is not related to the actual long-term cost of sustainable service provision which in turn results in inadequate levels of maintenance and investment in infrastructure. Furthermore, the annual budget allocation is often determined considering only maintenance costs rather than considering the life cycle cost of the infrastructure in which other events such as rehabilitation and modernisation would also occur. This traditional approach is termed *"input driven"* budgetary management.

The key feature of service oriented management is the *"output driven"* nature of the budgetary process. The budgets of the authority must be based on its short and long term needs. Hence the financial management strategy of the organisation must be focused on the identification of resources required for the sustainable provision of the agreed level of service. Under this concept, all the organisation's plans must be evaluated in relation to the desired level of service. This implies that the financial consequences of different options for service levels should be understood and quantified. In this context, asset management programmes become an essential tools to:

♦ assist the organisation in conducting an extensive forward planning of the changes required in the infrastructure to improve both the service provided and reduce the overall cost of service provision;

♦ assist the organisation in defining the medium and long-term financial consequences of these changes

6.2   PRINCIPLES OF ASSET MANAGEMENT

6.2.1   Defining Asset Management

The application of asset management principles to publicly or privately owned irrigation and drainage infrastructure is a relatively new concept. Traditionally, the investment made in irrigation and drainage infrastructure by government was focussed primarily on the cost planning and constructing the infrastructure with little attention to the consumption of assets during their economic life. However, the management of infrastructure comprises several other types of events including maintenance, rehabilitation (replacement), modernisation or the implementation of new technology, retirement and disposal of assets. All these events have specific costs which form part of the overall cost of providing a sustainable service. Failing to consider all the events occurring during the life cycle of the infrastructure precludes the identification of all the costs involved in provision of service. A program that takes into account all these infrastructure events is termed *Asset Management Program*. Such a program allows management to establish the cost of providing a level of service in a sustainable manner based on well defined and measurable performance parameters. Thus asset management must be seen integrally in its relation with the provision of different *Levels of Service* and *Cost of Service* (Fig 6.1).

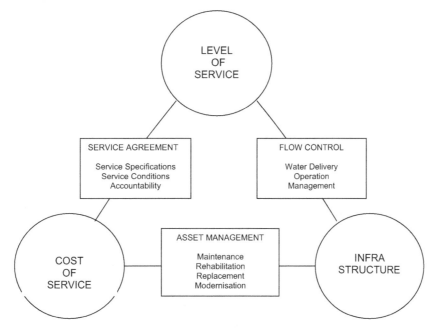

Figure 6.1   Asset management in service oriented management

Hofwegen and Malano (1997) defined an asset management program for irrigation and drainage services as "...*A plan for the creation or acquisition, maintenance, operation, replacement, modernisation and disposal of irrigation and drainage assets to provide an agreed level of service in the most cost-effective and sustainable manner*". Sustainability and cost effectiveness are the two critical aspects addressed by an asset management program. By determining the long-term actual cost of owning and operating the hydraulic infrastructure, management can become aware of the level of revenues necessary to guarantee its long-term sustainability, or of the eventual inadequacy of current funding. Box 6.1 provides definitions of the activities that take place during the life of assets as used in this monograph.

6.2.2 Basic Asset Management Concepts

Types of assets

In standard accounting practice assets are defined as that which is owned and liabilities as that which is owed. This definition embodies both types of assets: current assets and fixed assets. While current assets are important from an accounting perspective, from the viewpoint of

---

Box 6.1. Definition of terms related to irrigation and drainage infrastructure activities

Maintenance
Process of keeping the irrigation and drainage infrastructure assets in good repair and working order, to fulfil the functions for which it was created.

Major repair
Process of replacing or repairing a significant component of an infrastructure asset.

Rehabilitation
Process of renovating an existing asset whose performance is failing to meet its original objective.

Replacement
Process whereby an existing asset that is failing to meet its original objective is replaced by a new asset of similar characteristics.

Modernisation
Process of upgrading an existing asset in order to meet enhanced technical capacity to meet level of service objectives.

Renewal
General term used to describe rehabilitation, replacement or modernisation events.

---

an asset management program for irrigation and drainage infrastructure the focus is entirely on fixed assets. These in turn can be further classified as static assets and dynamic assets (Figure 6.2).

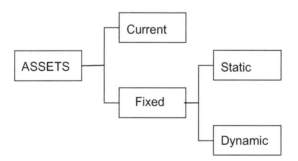

Figure 6.2 Classification of irrigation and drainage assets.

Static or passive assets are those fixed assets that do not have significant moving parts subject to wear and tear and periodic replacement. These include assets such as pipelines, canals, bridges, roads, etc. Dynamic assets involve significant moving parts such as pumps and plant and equipment, which require periodic monitoring and replacement.

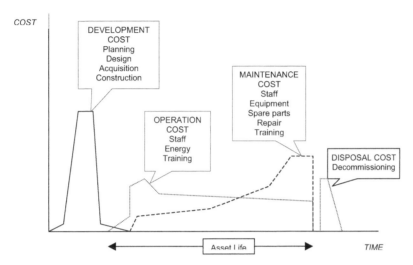

Figure.6.3 Typical life cycle cost profile of an irrigation and drainage assets

Life Cycle

The life cycle of a structure or a fixed asset is presented in Figure 6.3. It starts with the planning, design and construction of the structure after which commissioning occurs and operation begins. The asset will then be operated and maintained during its economic life. Because of wear and tear its condition will gradually deteriorate and may need to be replaced, rehabilitated or modernised. These events are generally referred to as renewals. If the asset is replaced or modernised, it must be removed and disposed before the renewal can occur. Figure 6.3 also depicts the relative cost associated with each phase of the life cycle. It is often assumed that the planning and creation of the infrastructure accounts for most of the life cycle cost. In reality, a detailed analysis of the operation and maintenance costs that occur over the entire life of assets shows that these are also very significant.

The period during which the asset is in operation and remains unchanged between planning and disposal is called the *life cycle*. The age of the asset is the number of years elapsed since commissioning.

The *economic life* of an asset is the time interval that minimises the asset's total equivalent annual costs to meet a specific function. Often, the economic life is shorter than the physical life as a particular asset can still remain in operation after the end of its economic life if the maintenance level is appropriate. This often implies an increased maintenance level, which may not make it the lowest cost option to perform the particular function.

The *useful life* of an asset is the actual time period that an asset is expected to be used. An asset's useful life is affected by many factors including for example the quality of construction and geological conditions for canal linings, capacity exceedance for pumping plants, etc. Unlike economic life which is determined by forward estimates based on previous experience with similar assets, the useful life requires to be updated by periodic condition assessments on the basis of the current condition of the asset.

The *residual life* is the expected remaining life of an asset at any point in time. Like with useful life, it varies as a result of periodic condition assessments throughout the life of the asset.

6.3   THE ASSET REGISTER

The register or inventory of assets forms the basis for the development of the asset management program and represents the single largest cost of implementing the asset management program. The asset register is a database of asset information. The database can be developed manually or electronically, however, the need for data manipulation later in the program makes the use of computer databases very desirable. The asset register is based on a survey of the asset base and its level of detail is determined by the aims of the asset management program. For a program aimed at valuation of the asset base for transfer of ownership (privatisation), sampling methods that rely on broad categorisations (bands) of assets according to type and condition may be adequate. These methods are based on representative samples of the various asset types which do not require detailed information on condition description (Metcalfe, 1991). Statistical techniques based on a subjective probability of occurrence of certain events such as replacements, retirement, etc. can assist in inferring the long term costs associated the future management of the assets.

However, development of an on-going asset management program for irrigation and drainage infrastructure requires a comprehensive survey of all individual assets. They must be initially inventoried according to specific condition rating criteria to enable detailed monitoring of the asset performance. All these data are entered into the asset register.

The design of the asset registry and the information to be included depends on the objectives set out for the asset management program. However, for any asset register there is a minimum requirement of data to be included in the database. Typically, these will include:

- *identification code:* Code systems vary in the level of detail involved. However, a unique code must be assigned to each asset that can be used throughout the agency for operational, accounting and financial purposes. A unique number will enable management to identify the asset including its hydraulic, operational and financial features so that a common database can be used for all the agency information technology needs;

- *type of structure and other characteristics*: Irrigation and drainage assets can be grouped in various ways. However, a common grouping is according to the type of structure, e.g. canal, cross regulator, offtake, pump, bridge, etc. A description of the main characteristics of the asset including location, size, geometry and construction material is necessary. Some of this information can be stored by using a classification code complemented by a narrative description and in some cases a photo of the asset;

♦ *asset age:* The date of the asset commissioning is necessary to carry out annuity calculations of asset depreciation and to create an 'age profile' of the entire asset stock. This information is more critical for old assets that may be approaching the end of their life and are in need of rehabilitation, replacement or major repair;

♦ *asset residual life:* The actual remaining life of the asset before replacement or rehabilitation is needed to determine the investment program required for maintaining the specified level of service. The residual life is usually associated with the condition and age of the asset. Residual life estimates are often subjective and also variable with the level of maintenance performed on the asset. Clear criteria must be developed and adequate training must be provided to asset surveyors before undertaking an asset survey. The residual life of the asset must be adjusted periodically according to the changes in asset condition;

♦ *asset condition:* The future actions to be taken on a particular asset depend primarily on its present condition and its ability to perform its specific function to meet the established level of service. The evaluation of the asset condition must be based on an objective inspection of the asset. An important consideration in the selection of the condition assessment is that the process and the level of sophistication, must respond to the needs of the organisation and its strategic goals in regard to the provision of service. The sophistication of the assessment criteria may range from a simple ranking to composite multi-dimensional criteria that may include a detailed assessment of each asset component as well as of the whole asset. The cost of the program is influenced by this level of detail. A simple asset condition rating criteria was employed in the ranking of channel and pump condition in the La Khe Irrigation Scheme in Vietnam (Box 6.2). This example illustrates the first of a two-stage process used to rate the conditions of assets in this scheme which involved the survey of irrigation and drainage channels, bridges, flumes, cross regulators, offtakes and pumps. All types of assets were ranked on a scale of 1 to 4 according to their condition;

♦ *replacement or reproduction cost:* This is the cost that will be incurred in replacing an existing asset at the end of its service life. If the asset will be modernised by using improved hydraulic control technology then the replacement cost must be considered. Otherwise, if a similar type of asset will be use in the replacement, for instance using identical hydraulic control technology, the reproduction cost of the asset is used.

♦ *written down value:* It is the residual value of the asset after depreciation. If substantial work has been performed to repair or modify the asset, the residual value and the cost of the repair will represent the new residual value. Asset valuation often must comply with accounting standards that are country specific and which may also specify the method of depreciation to be used. In the absence of any repair, the value of the asset is calculated automatically on the basis of the asset age and depreciation method;

♦ *historic information:* The history of the asset provides the best base on which to make decisions concerning the future of the asset. Of particular importance, is information on maintenance, major failures and repairs, modifications, and specific constraints to deliver the specified level of service observed during the monitoring of level of service provision;

- *general comments:* There are often particular features of an asset that do not occur except in individual cases but which are useful to the asset manager. Provisions must be made to record these occurrences in textual form as part of the database.

Box 6.2 Condition rating criteria for channels and pumps for the La Khe irrigation scheme Vietnam (Malano *et al*, 1997)

Condition rating for channel embankment			
1 *Excellent*	*2* *Minor defects*	*3* *Needs major repair*	*4* *Needs renewal*
◆ Excellent condition ◆ Recently constructed/rehabilitated ◆ No observable leakage or seepage	◆ Some cross section reduction ◆ No significant capacity constraints	◆ Cross section significantly reduced ◆ Leakage and seepage through banks ◆ Significant capacity constraints	◆ Cross section severely reduced ◆ Severe leakage and seepage ◆ Potential for bank collapse

Condition rating for pumps			
1 *Excellent*	*2* *Minor defects*	*3* *Needs major repair*	*4* *Needs renewal*
◆ Recently installed, repaired or replaced	◆ Routine maintenance and minor fitting required ◆ Operation not affected	◆ Replacement of parts required ◆ Load capacity and operation affected	◆ Major overhaul required ◆ Loss of capacity and potential for major breakdown of pump, motor or electrical equipment

6.4   ASSET MANAGEMENT FUNCTIONS

An asset management program is characterised by a strategic and integrated analysis of the life cycle of the infrastructure to determine the actual cost of owning and operating the infrastructure assets. It aims at obtaining the most cost-effective long-term strategy to deliver the specified level of service. In other words, as the definition implies, the asset management program must be able to provide a clear picture to the organisation and users of the financial implications of providing the agreed level of service. It does so by considering the long-term cost of owning and operating the irrigation and drainage assets.

Figure 6.4 illustrates the management functions that form part of an asset management program. These functions cover the entire life cycle of the assets and constitute a comprehensive process of analysing all the structural and non-structural options that are available to provide and sustain the agreed level of service. This process differs slightly depending whether we are dealing with the creation of new assets, e.g. the development of a new irrigation scheme, or the management of an existing scheme to meet an agreed level of service. Creation or acquisition of new assets may take place both in new and existing schemes. In existing schemes, however, they usually form part of an augmentation process designed to provide an enhanced level of service. Whether for new or existing schemes, the asset management program is an integral part of, and must be consistent with the goals of the strategic management plan of the organisation. The set of functions the program must perform is essentially similar for irrigation and drainage infrastructure.

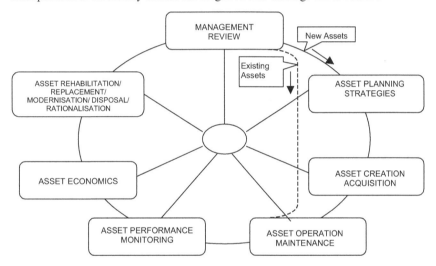

Figure 6.4 Functional elements of an asset management program

6.4.1   New Assets

New assets are created or acquired either for development or expansion of new schemes or augmentation of existing ones aimed at improving the level of service. This can only be justified after a thorough management review of the whole system to identify if there is a clear need expressed by:

- ♦ the authority relating to reducing internal management cost;
- ♦ present or potential clients relating to adjustments in the service level, service cost or expansion of the service area;
- ♦ public interest in general through government or river basin authorities to meet drainage effluent standards to minimise impacts on receiving waters.

Following this review and having identified a need for new assets, strategies are developed for the planning and design of these new assets. A comprehensive evaluation is carried out to explore all the alternatives available to meet the new service needs and to identify the consequences for the level of service and the cost of service provision. These asset planning strategies could even consider utilisation of second-hand assets, especially for mechanical and electrical equipment such as pumps and related electrical plant. It is important that in this evaluation all life cycle costs are clearly identified including planning, design, construction or acquisition, operation and maintenance and disposal.

The information related to any new asset should be well documented and incorporated into the organisation's asset register as soon as the asset has been commissioned. This should include all the specifications of the asset and details of the asset's operation. The information contained in the register will then be used for decision making during the various stages of the asset management programme.

6.4.2   Asset Operation and Maintenance

Operation and maintenance expenses often represent a large proportion of the life cycle costs of assets. These are higher for the so-called dynamic assets such as pumps, plant and equipment, vehicles, etc. But also for passive assets such as canals and control structures, maintenance plays a critical role in determining the economic life of these structures. Poor maintenance standards have an impact on the life as well as the ability of assets to perform their intended function. However, maintenance is highly subjective and what is considered good or poor standard varies in relation to the type of asset and the relative importance of the various costs associated with maintenance work (Niebel, 1985). The main costs relevant to irrigation and drainage infrastructure are:

+ *Direct costs:* Costs required to keep the asset operable. These include periodic inspection, preventative maintenance, repair costs, etc.

+ *Standby costs:* The cost of operating and maintaining standby equipment needed to be put in operation when primary facilities are inoperable, eg. drainage pumping units that operate continuously to control water level in drainage canals.

+ *Lost production costs:* Losses due to downtime of primary equipment when no standby equipment is available thus preventing the organisation from delivering service, eg. suspension of service due to shutdown of pumping station or pumping units for repairs.

+ *Degradation costs:* Costs occurring in deterioration during the life span of assets resulting from inadequate and/or substandard maintenance.

All these costs have specific impacts on the overall cost of maintenance and vary with the amount of maintenance performed on assets; and more maintenance does not necessarily translate into better maintenance. For instance, the cost of lost production due to downtime of a pumping unit may reduce initially if the amount of maintenance increases because of improved reliability. Nevertheless, if the amount of maintenance keeps increasing the cost of maintenance may begin to offset and more than compensate for the savings due to improved reliability.

Operation of assets in the context of asset management is defined as *the manipulation of assets with the specific purpose of performing the asset design function to deliver the specified level of irrigation and drainage service.* The efficiency with which the asset is operated can have a significant impact on the overall cost of operation. A detailed analysis is required of both the history of assets' operation and maintenance and the future operational requirements to ascertain the efficiency of asset operation. This would consider factors such as the actual performance of the asset in relation to the design or commissioning tests, the overall operation cost, and the capacity at which the asset is currently operating in relation to the design assumptions.

Changing operating procedures can sometimes reduce the cost of asset operation. For instance, changes in the operation of pump operating hours may reduce the cost of energy where differential pricing of electricity is applied according to the time of usage. Improved sediment management upstream of the pump intake may increase the life of pumps. However, this type of decision making must rely on comprehensive information of the historical performance of assets. The historical overview should contain both the costs of operation and maintenance, failure record and its relation to maintenance levels. Based on this, overall targets for future operation and maintenance can be determined. These include the probability of failure, allocation of O&M funds, and whether changing operational requirements can extend the life, reduce the cost of operation and improve the performance of the asset.

Maintenance records are critical to the ability of managers to carry out an operation and maintenance analysis. By integrating maintenance records with the asset database, management can quickly access information on the performance of individual assets. Maintenance practices must be reviewed periodically to ensure that they are in line with the functions and purpose of the asset and to analyse the cost of maintenance needed for meeting performance requirement against replacement costs. Computerised records of maintenance and related activities are required for effective work control, inventory control for spare parts and cost, performance monitoring and reporting.

6.4.3   Asset Condition and Performance Monitoring

Each asset in the irrigation and drainage system is designed to perform certain functions. The capability of structures to perform these functions reduces over time due to age and wear and tear. To maintain the provision of service, detailed knowledge of the agency's asset stock and performance is required. Often, the failure by irrigation and drainage organisations to maintain up-to-date knowledge of their asset holdings and asset condition and performance has led to inadequate maintenance or untimely replacements. The consequent failure of assets often leaves the agency with no option other than that of rehabilitating or replacing the asset. This entails the critical notion that condition monitoring is an on-going task that enables the asset manager to make decisions concerning the future of each asset in relation to the provision of the intended level of service.

Asset Condition and Serviceability

Irrigation and drainage assets age at different rates throughout the system due to varying operating conditions and maintenance. This differential ageing can occur both within and between asset categories. The condition profile of the asset base can be developed from the asset register to provide a snapshot of the overall distribution of asset quality.

An example of such profile for the La Khe irrigation system in Vietnam is shown in figure 6.5. The condition profile also provides an indirect indication of the investment priorities and timing. A more detailed description of the investment plan and strategies is discussed in Section 6.5.4.

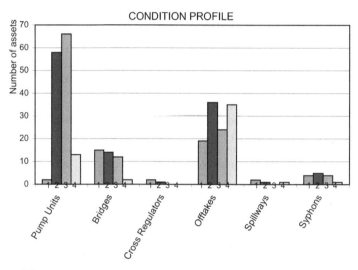

Figure 6.5. Condition profile of various asset categories at La Khe irrigation system, Vietnam (Malano *et al*, 1997)

A good physical condition of the asset does not automatically mean that it can perform its function adequately to deliver a specific service. For instance, a control structure with visible signs of ageing may still be adequate to control the water level if all its components are still in operating condition; on the other hand a new structure of inadequate dimensions or inadequate gate lifting devices would constrain the level service provision. Therefore structures also have to be evaluated based on their ability to provide the required service. This is often referred to as *serviceability*.

The concept of asset serviceability is based on the ability of the asset to deliver a specific level of service. This concept was first developed in relation to the management of urban utility assets in the United Kingdom. Water companies realised that improving the condition of a particular asset did not necessarily affect its ability to deliver a specific service, and therefore did not alter the overall system performance (Banyard and Bostock, 1998). This was expressed in the form of a "Serviceability Matrix" whereby the level of serviceability is based on a number of physical and service parameters. Burton and Hall (1998) have attempted to apply this concept to irrigation assets. These authors devised a serviceability matrix based on quality criteria for irrigation service from the perspective of the clients (farmers) such as adequacy, timeliness, equity and reliability of supply and infrastructure performance indices such as hydraulic performance, condition and importance of assets. The use of service performance indices to determine the serviceability of assets links asset performance directly to service provision.

Management of the irrigation infrastructure and actual asset physical condition play an equally important role in the ability to meet service specifications as discussed in Chapter 5. The specific contribution of management and physical condition to the delivery of a specific level of service is often difficult to segregate. However, it is important that the criteria used to assess the performance of assets also include sufficient information on its management. Selecting appropriate types of indicators to measure both the physical and the management aspects of asset performance would enable managers to make strategic decisions based on the cost of operation and of maintenance of each asset. Situations can arise where existing irrigation and drainage assets can meet serviceability criteria albeit at a very high cost of maintenance and repairs. Likewise the service obtained from assets in good condition may be low due to poor operation practices. Operational analysis is a function that must be carried out as part of the asset management program in which operation and maintenance practices of assets must be analysed to determine their current costs and potential for reduction through changes in current practices or renewal of the asset.

Asset Performance Monitoring

Asset performance must be assessed against the function of each specific asset in delivering the agreed level of service. Measures of service delivery performance can provide an indication of the capacity of the asset and the concurrent management to provide the required service now and in future. There are however intrinsic asset properties that must be measured and used to evaluate asset performance. Typically these indicators express the physical capacity of the asset or the financial implications of operating and maintaining the asset and meeting safety standards. Examples of these are:

- maintenance and operation costs;
- frequency of breakdown;
- downtime;
- capacity utilisation;
- integrity of components.

Historic data on asset maintenance, breakdowns and downtime are essential to evaluate the reliability of assets. This information, which usually resides in the asset register, enables the asset manager to carry out the relevant maintenance and renewal analysis for individual assets.

Often, cost reductions can be achieved at the expense of reducing the efficiency with which the asset can fulfil its function. In considering the effectiveness with which assets fulfil their function, the asset manager must carry out a careful analysis of all the management (software) and physical (hardware) costs pertaining to each asset. The governing criterion to establish any cost comparison must be the provision of a well-defined level of service. For instance, both flat sluice gates and automatic gates can provide upstream control in a main irrigation canal. However, if the level of service provided includes on-request deliveries with the option of selecting discharge, duration and timing of the delivery by the irrigator, then automatic gates are more capable of coping with frequent water level variations. Manually operated gates would have a much higher operation cost due to increased operational inputs in the form of staff time and skill levels.

Complete condition monitoring checklists must be drawn up to ensure that asset management staff applies consistent performance evaluation criteria. Additional decision support tools can be developed to assist the asset manager in carrying performance and replacement analysis. In preparing such checklists it should be realised that the performance targets can change with time due to changes in the agriculture sector. For instance irrigation assets that were designed to supply water to rice mono-culture may no longer be adequate to supply water to a more diversified cropping pattern. A typical example of asset under-utilisation is that farmers prefer not to irrigate at night. If the irrigation authority responds to their preference, the capacity of the supply system would have to be enlarged to satisfy the irrigation demands during daytime leading to an under-utilisation of the available capacity during night-time. An alternative solution would be introducing on-farm storage to hold the system's night flows. Pumping capacity for drainage in the humid tropics is a typical example of the capacity of assets to meet the required demand. A decision to increase the level of protection against prolonged flooding entails a higher cost of service. The asset management program must evaluate all the alternatives available and select that with the lowest cost, or with the best result.

6.4.4 Asset Accounting and Economics

The calculation of infrastructure costs requires the identification of all the costs of the asset that accrue during its life cycle. As described earlier, an asset goes through several major stages in its life cycle: planning, design, construction or acquisition, operation and maintenance, rehabilitation, modernisation and disposal. Design and selection decisions made during the planning stage will have a greater impact on the life cycle costs than do decisions made at a later stage. However, decisions should consider costs in conjunction with the efficiency and effectiveness of the asset itself.

An accounting and economic analysis provides the basis for the identification of all the recurrent as well as non-recurrent cost in the life cycle of the asset. Accounting for all the costs provides the basis for managers to formulate long-term strategies for maintenance, rehabilitation and modernisation of the infrastructure. Costing of assets must therefore be carried out on the basis of individual assets. By doing so, management can plan the timing of interventions to stage and smooth out events involving significant outlay of funds for rehabilitation, replacement and modernisation of assets. In addition, organisations may face specific accounting legal requirements such as accrual accounting[11] in which both cash and non-cash costs must be considered.

Life Cycle Costing of Assets

The recognition of all the life cycle costs associated with an asset is critical to the formulation of the cost analysis. The cost items that arise during the life cycle of irrigation and drainage assets are listed below:

- feasibility, planning and design cost (investment);

- construction or acquisition costs (investment);

---

[11] Accrual accounting recognises revenues and expenses over the accounting period irrespective of whether cash is paid or received. It includes items such as cost of consuming assets (depreciation), accruing employee entitlements, etc.

- operations and maintenance costs (recurrent);
- rehabilitation and modernisation costs (investment);
- disposal cost (investment).

The relative magnitude of these cost components varies with different assets. The relative importance of various costs also varies throughout the life of an asset. Although a generalisation is difficult, planning, construction, rehabilitation or and modernisation costs usually occur over a shorter period of time than operation and maintenance costs. Box 6.1 illustrates the main activities and associated costs that occur over the life of an asset which constitute the overall life cycle cost profile. Most of the cost items identified in the list above are self-explanatory.

While the list of costs presented above represents the actual cost items associated with the various stages in the life of an asset, other forms of expressing asset costs are required to satisfy accounting requirements and carry out replacement analysis. Among these are:

- depreciation;
- rate of return on investment; and
- risk cost.

Depreciation

The financial analysis of irrigation and drainage services must take into account the full cost of depreciation for the investment components of the asset cost. All assets that form part of the irrigation and drainage infrastructure have a finite life span. The use of assets to provide irrigation and drainage services will reduce their value with the passage of time. This is reflected in the rate at which capital assets are consumed in the process of providing a service. The lessening of the asset value is reflected through its depreciation. Thuesen & Fabrycky (1964) recognise two types of depreciation: (a) physical depreciation and (b) functional depreciation or obsolescence.

Physical depreciation results from the physical impairment of the asset by way of deterioration due to corrosion, chemical and biological action of structures, etc; and, wear and tear from use such as abrasion of pump impellers, bearings, etc. *Functional depreciation* due to obsolescence arises not from the ability of an asset to perform its intended function, but from the emergence of new technology that can perform that particular function more effectively and efficiently.

Obsolescence can also be due to a change in the type of functions that the asset must perform. As an example, consider an irrigation scheme with a manually operated upstream control system that is supplying water under an imposed arrangement with fixed rates and duration. Notwithstanding that the infrastructure may still be in good condition, an upgrade in the level of service in which water will be supplied more flexibly may not be possible unless local or centralised automatic control is implemented. In this case, the old assets will be replaced and modernised to meet the requirements for the new level of service despite the fact that their physical condition is still adequate.

The financial analysis of irrigation and drainage services must take into account the full cost of depreciation. This cost reflects the rate at which capital assets are consumed in the process of providing a service. There are several approaches that can be used to calculate the rate of asset depreciation. The more common methods are (a) the straight-line method, (b) the sinking-fund method, and (c) the fixed percentage method[12]. To calculate the rate of depreciation of irrigation and drainage assets the straight-line method of depreciation is generally used. Other methods differ in the rate at which assets are depreciated at different stages of their life. A detailed discussion of the various depreciation methods is beyond the scope of this monograph[13].

The calculation of asset depreciation is based on the value of the asset, the estimated salvage value[14] and the service life (useful life) of the asset. The initial cost of the asset is known with certainty whereas the salvage value and the service life of the asset must be estimated, as these are events that occur in the future. However, for most of the irrigation and drainage assets the salvage value can be considered negligible except for items such as pumps and electric equipment when they may be sold as second hand equipment.

Assets must be assigned an initial value to enable their residual (written down) value to be accurately calculated. There are main approaches that can be used in the valuation of irrigation and drainage assets: (a) *replacement cost,* and (b) *reproduction cost.* By using the replacement cost criterion, the asset valuation reflects the cost of using new technology in the new asset. Reproduction cost is used when the existing asset will be replaced with the same type of asset. The situation of replacement cost often arises when an upgrade in the level of service to be provided by the irrigation agency involves substantial modernisation of the irrigation or drainage infrastructure. Consider for example the implementation of automatic downstream control where canal grades would have to be changed and automatic control gates would need to be installed. The valuation of assets would have to consider the value of the assets with which the existing ones would be replaced in order to reflect the actual cost of the upgrade in the level of service.

The service life assigned to an asset is usually based on experience and is essentially a matter of judgement. Various engineering textbooks provide indicative values of irrigation and drainage equipment and appurtenant structures (Jensen, 1980). It must be borne in mind, however, that because of the uncertainty involved in the service life of an asset, the written-down value often does not reflect the actual value of the asset regardless of the depreciation formula that is used. Furthermore, the rate of decay of an asset sometimes will vary throughout its life as a result of varying levels of maintenance and when periodic major maintenance work is carried out. This is critically important for *infrastructure assets* where lower than required expenditure on maintenance results in loss of future service ability. In these cases, depreciation formulae do not reflect the actual decline in the value of the asset often overstating the actual residual value and understating the future asset life. It is therefore important to emphasise that only experienced personnel should carry out the condition assessment of existing assets. With the implementation of the asset management program, the agency will be able in the medium term to make this type of

---

[12] Straight-line method: The value of the asset decreases at a constant rate. Sinking-fund method: The value of the asset decreases at an increasing rate. Fixed percentage method: The value of the asset decreases at a decreasing rate.

[13] There are a number of Engineering Economy textbooks in which the interested reader can find a detailed discussion on this topic.

[14] Salvage value is the residual value of the asset at the end of its economic or useful life. It is usually negligible for irrigation and drainage assets.

assessment more objectively on the basis of the information stored in the asset database. This is actually considered an additional benefit of the asset management program.

An alternative approach to assessing the actual depreciation of an asset is by using *Condition Based Depreciation (CBD)*. This method assumes that the asset value is determined by its deviation from the "as new" condition. For instance, if a canal cross regulator has a replacement cost of $150,000 and the cost of bringing the asset to "as new" working order is $25,000, then its written-down value is $125,000. If the process is repeated at periodic intervals, the change in the residual value can then be ascertained as illustrated in Figure 6.5.

The residual life of assets must be adjusted on an on-going basis, as more information about the asset condition and performance becomes available. As a result of reassessing the residual life, adjustments in the asset depreciation may have to be made. A residual life shorter than previously estimated represents a greater loss of service potential, which translates into greater depreciation charges. Conversely, if the residual life is greater than previously estimated, depreciation charges can be reduced. To enable the asset manager to update the asset valuation continuously, asset information must be stored in a comprehensive database that can be continuously updated to reflect the actual asset condition.

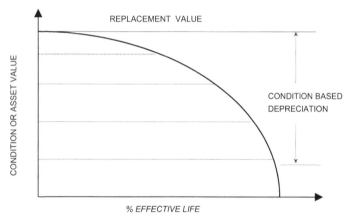

Figure 6.6   Illustration of condition based depreciation (After IMEA, 1994)

Irrigation and drainage organisations often embark in developing an asset management program after many years of operation without accounting for the cost of assets. In such cases, the use of straight depreciation techniques is not adequate to ensure the long-term financial viability of the organisation since part of the asset base has already been consumed. An alternative *renewals based* approach to the calculation of costs is required to ensure that the appropriate annuity is set aside for the capital expenditure needed over the planning period. Under this approach, a net present value annuity is accumulated to cover the future investment in asset renewals. This is a common approach used by irrigation and drainage agencies that operate assets originally developed and subsidised by Governments with no provisions for capital recovery.

Two illustrations of this approach to financing asset renewals are the Societes d'Amenagement Regional (SARs) in France, and Irrigation Districts in South-East Australia. The SARs were created with a specific mandate to design, implement and operate the water resources infrastructure for irrigation, urban and industrial water supply. The SARs have a mandate for a period of 75 years at the end of which they must ensure that the infrastructure is in good condition. The asset renewals required to meet this obligation must be funded their own service revenues. (Tardieu and Plantey 1999)

The development of Irrigation Districts in South East Australia early this century, especially in the Murray-Darling basin was also heavily subsidised by the States and Commonwealth Governments. The initial development and subsequent operation of the infrastructure was carried out by State Government Departments or Statutory authorities. More recent changes in State Government policy in New South Wales and Victoria discontinued the policy of subsidies to irrigation authorities forcing the full recovery of the cost of asset renewals.

This approach yields higher cost of services than the straight depreciation or CBD in most cases because of the shorter time horizon available to recover the renewal cost. It is however a more realistic alternative especially when no provisions were made historically for asset renewals prior to the implementation of the asset management program. The continuous application of this method will guarantee the ability to sustain the provision of the selected level of service.

Investment Profile

A common feature of irrigation and drainage assets is the "lumpiness" of the investment required to replace various assets. Often, large groups of assets such as offtakes, regulators and pumps must be replaced together because of similar decay rate in condition. The asset replacement profile is based on the rating of asset condition. Residual life must be ascribed to individual assets on the basis of the current condition. Assets with the same residual life are then aggregated over time to yield the overall investment required (Box 6.3). The residual life is estimated assuming a certain on-going level of maintenance for the rest of the asset life. Nevertheless, maintenance and other factors that determine the productive life of the assets may change over time altering the original assumed residual life. Use of expert advice and use of *value analysis* (see section 6.4.5) are often necessary to produce more accurate and objective estimates of the residual life and alternative choices for renewal or upgrade of assets. One of the main benefits of a life-cycle approach to management of irrigation and drainage assets is that it enables the authority to develop long-term investment strategies and to cope with the "lumpiness" of investment requirement.

Rate of Return

In addition to making allowance to cover the cost of asset consumption through a depreciation charge, the infrastructure investment may be expected to yield a certain return on the residual value of the asset. This rationale is based on principles of economic efficiency, and it is assumed to account for the opportunity cost of the capital invested in infrastructure development.

Irrigation and drainage policy often dictates whether irrigation organisations will expect a rate of return on the infrastructure investment. Government investment in irrigation and drainage infrastructure often does not include a positive rate of return as part of the overall cost of service provision. The rationale behind it is that Governments in general do not seek a return from this type of investment. For private irrigation and drainage

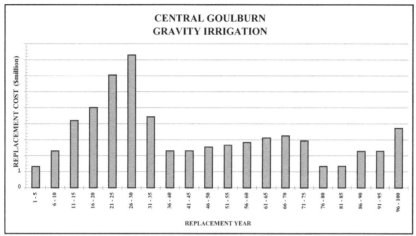

Figure 6.7   Investment profile for asset replacement for Central Goulburn Gravity Irrigation System, Australia (Moorhouse, 1998)

infrastructure in which the users themselves are the only stakeholders, there are no obvious incentives to consider a return on investment charge.

Asset Management and Service Cost

The ultimate outcome of an asset management program is the information needed to determine the cost of sustainable service provision. This is the integration of all the infrastructure and other operational resources such as personnel, financial and insurance costs and power use that the organisation incur in the provision of services (Figure 6. 4).

It is recognised that the cost of service provision may differ from the price of service (tariff). Government policy may be designed to provide irrigation and drainage services below cost with variable levels of cost recovery. The gap between actual cost and price of service provision can only be determined if a full cost analysis based on a long-term asset management strategy is developed. This gap must be made up by subsidies to the irrigation and drainage sector; otherwise the funding shortfall will be reflected in lack of infrastructure sustainability caused by either under-spending in maintenance or lack of depreciation provisions for the normal asset consumption or a combination of both.

If the irrigation and drainage organisation provides only a single service, then all costs must be assigned to that particular service. However, many organisations provide multiple services. While it is possible that cross subsidies may occur between different services within the organisation, it is imperative that the cost of provision of individual services, e.g. irrigation, drainage, flood control, are clearly delineated. This is necessary to analyse the performance of each service activity and recognise areas of potential productivity increases. Activity based accounting is well suited to assist management with this task.

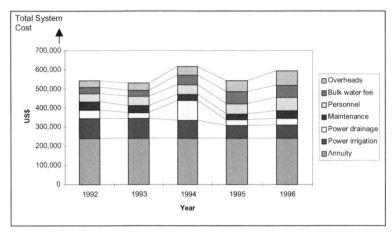

Figure 6.8   Cost of service provision for La Khe irrigation scheme, Vietnam

6.4.5   Asset Audit and Renewal

An asset management program deals with all the activities that occur during the life cycle of assets. In addition to maintenance activities, the program must include other events such as rehabilitation, replacement, modernisation and disposal of assets to account for their financial implications. Typically, most of these events constitute comparatively large investments for which the program must make adequate financial provisions.

As pointed out above, a salient feature of irrigation and drainage is the "lumpiness" of investment resulting from concurrent renewal of many assets of the same category or individual assets of high value (Figure 6.7). It is often not practical or realistic to lay out a financial strategy that encompass the entire useful life of all assets (useful life of dams may exceed 100 years). Shorter rolling investment programs (15 - 20 years) can be established to smooth-out the renewal outlays. Appropriate financial management arrangements to build up the required depreciation reserve must be put in place to make appropriate provisions for renewals.

At times of considering any renewal action, the current assets must be evaluated against the range of renewal options - structural and management-. Asset managers are often confronted with decisions such as escalating cost of maintenance versus rehabilitation, rehabilitation versus modernisation, amount of capital investment required and when this investment will be needed. The evaluation is based on the comparison of the economic future of the present asset and the alternative renewal measures.   The causes of performance shortfall must be ascertained to enable determination of the cost of sustaining the present assets. Irrigation and drainage assets can fall short of delivering the required service due to several reasons including:

- *inappropriate design or selection of hydraulic control technology:* Proper hydraulic design or selection of irrigation and drainage assets determines the ability to perform their intended function. For instance, control structures incapable of handling the required discharge or maintain the required flow levels, pumping stations unable to deliver the required peak discharge to meet crop water demand are illustrations of inappropriate design. A certain level of service could be provided by different types of hydraulic control technology and the associated management. Less automated systems require greater management input to achieve a given level of service, e.g. manual upstream control can be used to deliver water on request with no constraint in duration, flow and frequency. The number of staff would be substantially reduced if the same level of service is to be provided by an automatic upstream control system although the level of skills required would be greater;

♦ *inappropriate construction:* Failure to meet construction standards is a common factor in new assets being unable to fulfil their specific function. For instance, inappropriate downstream protection of flow control structures could lead to reducing discharge to prevent scouring. Such a deficiency would also reduce the useful life of the asset;

♦ *condition (wear and tear):* Poor asset condition results in increased maintenance costs. The history of maintenance of the asset will reveal the changes in maintenance requirement over time and show the trend in future maintenance cost for the asset. Depending on the relative importance of the asset, the cost of risk[15] attached to the asset could be equal or higher than the cost of maintaining the asset;

♦ *obsolescence:* The asset condition may still be adequate but the technology used to control the operation may have been superseded by new and more efficient technology, e.g. locally manual operated pump switching replaced by remote control microwave technology to control and operate pumping stations. In some cases, lack of spare parts due to changes in design may justify change of equipment despite the old equipment being still operable;

♦ *structural integrity (safety):* Certain assets may have reached the end of their useful lives and are in danger of catastrophic failure with potentially severe consequences, e.g. step down transformers and switching equipment for large pumping stations that fail at the time that their operation is required for flood control;

♦ *insufficient capacity due to changed requirements:* Change in agricultural systems over the life of the infrastructure may sometimes render the assets incapable of meeting the service requirements, e.g. conveyance systems designed for supplementary irrigation of rice-only may not be able to support a more diversified and modern agricultural system;

♦ *vandalism:* Intentional damage to irrigation and drainage assets is common in some irrigation and drainage systems. The effect of vandalism depends on the type of damage; but it is particularly severe in systems that rely on sophisticated equipment for flow control such as sensors, telemetry devices and actuators. Users

---

[15]. Cost of Risk = Probability of occurrence of failure x Financial damage due to failure

also may intentionally damage irrigation structures in an effort to obtain additional water supply rendering the asset unable to function.

Risk Assessment and Renewal Decision-Making

Asset renewal is aimed at expanding existing service capacity, introduce new water control technology or replace assets that have reached the end of their useful life. The cost associated with the risk of failure is related to the condition of structures. Ageing structures in poor maintenance pose a greater risk of failure than structures in good conditions. Sometimes it may be desirable or necessary to postpone renewal thus increasing the greater risk of failure. A simple approach often used to replace existing assets is based on the notion of economic life. In reality, assets may reach the end of their useful life before or after its economic life is over. If the performance of an asset is unsatisfactory before the expiry of its economic life and replacement is delayed until the end of its economic life, the risk of failing to provide service, or damage resulting from structural failure is increased. Conversely, replacement of assets sooner than achieving economic life would incur in waste of resources. The level of risk associated with assets is also dependent on the spatial location of assets. For instance, the level of risk from failure of headworks assets is greater than that of assets in the lower system in the same condition.

The type and magnitude of risk associated with irrigation and drainage structures varies widely depending on the size and function of the structure. For instance, failure of a large regulation dam could cause loss of lives and loss of economic benefits from irrigation and other economic activities whereas failure of a small irrigation control structure may only have an economic impact on the area irrigated under its command. The risk costs associated with common irrigation and drainage assets are primarily related to the economic losses, e.g. crop losses, soil losses, etc; caused by inadequate water supply or inundation.

The level of risk from irrigation and drainage assets can be *managed* by altering the condition of the asset either through maintenance or major repairs. This process requires firstly identifying the potential problems associated with a particular asset, e.g. "What might go wrong?", and secondly produce alternative actions to reduce or eliminate the level of risk. This process of decision making on accepting, reducing or eliminating risks based on a qualitative and quantitative analysis is called *risk management.*

The qualitative analysis of risks deals with the cause and the consequences of probable failure. This analysis must follow a systematic approach that requires expert personnel to identify problems that are likely to occur with various assets. The process often produces a large number of potential risk factors, which must be evaluated and ranked. Quantitative and qualitative techniques can be used for this assessment for which computer software is available.

Quantitative assessment and classification of risk is based on the likelihood of occurrence and the impact of occurrence in financial terms. In principle, it is possible to relate the condition of the structure with the cost of risk. This process is called *risk assessment.* The determination of the probability of failure requires knowledge on the failure behaviour of identical or similar structures. Historical data series on failures will have to be translated into probability curves for failure of specific assets. Use of such curves requires compliance with the standard (design, construction etc.) for which these have been developed.

The process of risk assessment has three main elements: (a) identification of hazards, (b) evaluation of risk exposure, and (c) evaluation of response alternatives. The identification of hazards consists of the various events that might occur with the structure given the condition it is in. This includes the description of a "candidate problem" and consequences associated with the occurrence of the problem. For example breaching of a canal embankment will cause flooding of adjoining areas and interruption of supply to downstream users; failure of a pumping unit will interrupt supply for the duration of downtime, or if used to provide drainage service it will cause additional flood damage.

The risk exposure of an asset due to each hazard is the quantitative description of the risk including the probability that such events will occur. Risk exposure may be quantified and expressed for the asset as a whole or for each part separately and then integrated into a composite value. Often, the information required for determining the risk exposure is not readily available before the implementation of an asset management program. Both hazard identification and risk exposure must be assessed by technical specialists using an appropriate appraisal technique such as value engineering for the allocation of risk estimates that are expressed as an annual probability.

Response alternatives are actions that can be taken to change the level of risk of assets. Evaluating response alternatives is aimed to answer the question "how safe is safe enough?" These alternatives could include doing nothing, rehabilitating or modernising the asset or modifying its operation. The evaluation must be made considering the *residual risks* after the implementation of the corrective measures and the cost associated with each alternative.

Renewal Decision Analysis

The evaluation of asset performance and renewal analysis must be based on objective and accurate information about the assets. The asset register contains information on the technical and financial performance of each asset. The process of making renewal decisions involves the consideration of both technical and financial factors. The main technical factors that must be considered are often related to the type of hydraulic infrastructure that is required to provide the specified level of service. At the time of selecting the best option for hydraulic control, a number of options must be considered that involve a range of technological and operational inputs. In addition to economic considerations, the selection process must pay sufficient attention to the socio-economic setting of the irrigation scheme when selecting the mode of flow control.

The first action consists of predicting the remaining life of the asset by predicting the future decay and the associated risk. This is often a difficult and subjective task unless sufficient information has been collected on modes of asset failures to provide a given function. The ability to predict the remaining life of assets improves during the asset management program as more information becomes available through the monitoring function. Once the risk levels and associated costs are predicted, various options for renewal must be evaluated by contrasting the current O&M cost of the asset and the cost of risk associated with keeping the asset.

The main aim of the renewal analysis is to identify and select the optimum course of action that can be taken for a particular asset. Any renewal decision involves optimising both technically and financially the selection of options available to correct the performance shortfall of the asset. These may include do-nothing, continue the same maintenance level, rehabilitate, modernise or others. The number of options left available may also reduce if the asset is nearing the end of its useful life and approaching total failure to perform its function.

The financial aspect of the analysis involves the assessment of the current cost and benefits of all the renewal options and performing a benefit/cost analysis or cost-effectiveness analysis for each option. The results are ranked on the basis of the benefit/cost ratio or cost-effectiveness to select the optimum option for the asset concerned.

Value Engineering in Renewal Analysis

Each asset has a specific function to play within the hydraulic control system. When the asset manager is confronted with the problem of making objective decisions about the future of a specific asset, (s)he must do so by taking into account the specific function of the asset, its ability to perform this function and its associated cost. *Value engineering (or value analysis)* is an objective and systematic tool that can be used in carrying out this analysis. The objective of value engineering is to generate information to enable total cost control anywhere within an asset's life cycle to perform a specific function. The main objective of value analysis is to achieve the total function of the system at the lowest overall cost. This can only be achieved if objective measures of the value of each function are available (Milne, 1972, Mudge, 1971 and Gaff, 1987)[16]. The American Society of Value Engineers defines value engineering as "*...a systematic application of recognised techniques which identify the function of a product or service, establish a monetary value for that function, and provide the necessary function reliably at the lowest overall cost.*" In irrigation and drainage systems, this technique can be applied to a number of decisions concerning the maintenance or renewal of existing assets since each asset has a specific function to perform. If an asset is not required to perform any specific function then the asset must be retired in order to reduce costs. With renewal, the asset manager is often confronted with alternative options as to whether to reproduce the same asset by rehabilitation or replacement, or to modernise the asset to provide a higher level of service or to provide the same service more effectively.

The most effective way to apply value analysis principles to irrigation and drainage assets is to take a *total system approach* to the assessment of asset functions and costs. For instance, the analysis of canal assets because of the hydraulic interactions between the structural elements is best performed by a simulation analysis, which includes all the control structures in the system. A multi-disciplinary team with diverse expertise relevant to the study can often provide a more comprehensive assessment of the various options available for asset renewal and the costs and benefits associated with each option. The main strength of value engineering is its ability to tackle often complex problem by considering a total system approach. Systems must be analysed and quantified considering all the life cycle costs involved in the provision of the service including the problems associated with each asset and their causes, the actual need for the functions performed by the asset and future economy, technology, social and environmental scenarios. In

---

[16]. These are comprehensive references where the interested reader can find this topic treated in more thoroughly.

examining the total system, it is necessary to separate the essential functions from the secondary or non-essential functions. Savings are normally associated with the elimination of the secondary functions. During this analysis, "add-on" costs such as excessive maintenance arising from a desire to extend the life of the asset or to improve its aesthetics would be detected and eliminated.

Teamwork is essential to value engineering. It must provide an appropriate environment where individual team members can be creative and thinking laterally in order to produce innovative solutions to existing asset problems. This approach helps create ideas and solutions that normally would not surface when people work individually.

When these principles are applied to irrigation and drainage assets, the main goal is to achieve the provision of a specific level of service at the lowest possible cost. The study must always start by focusing on the entire service activity as this constitutes the overall main function of the asset system followed by a thorough examination of each asset for its need and appropriateness to perform the required service. All the analysis techniques must be aimed at identifying unnecessary costs and removing obstacles for better value.

6.5 DEVELOPMENT AND IMPLEMENTATION OF THE ASSET MANAGEMENT PROGRAM

6.5.1 Organisational Aspects

The asset management program must be fully integrated into the overall strategic planning of the organisation to deliver the agreed or declared level of service. Its importance must be reflected in the organisation's goals and objectives and must become one of the several plans that form part of the overall strategic planning. Consistent with the goals of the strategic planning, the irrigation and drainage organisation must be fully committed to the development and implementation of the asset management program.

The formulation of the program itself involves two main phases: preparatory phase, and implementation phase. The preliminary phase consists of setting up the organisational structure required to formulate and implement the program. It is desirable to establish an asset management unit with the adequate infrastructure and human resources with sufficient technical expertise. This unit, which should be headed up by an asset manager, will have the responsibility for establishing and maintaining the asset management program within the organisation.

During this preparatory phase, an assessment of the existing operation and maintenance budget, existing work programmes and relevant information on current practices is collected and collated. A preliminary assessment of the quality of asset data, its accuracy and availability and the extent of the existing information systems is also conducted. Planning for integration of the asset management program with the rest of the organisation plans must take place during this initial phase in order to avoid the program to become an "island" within the organisation. Lack of integration with other related units especially financial and management information systems can greatly diminish the effectiveness of the program. A later integration usually becomes substantially more expensive to achieve; especially when incompatible information systems must be replaced to attain the necessary integration.

From the beginning it is important to develop ownership of the program across the organisation. It is imperative that the concepts and principles of the asset management program are explained to managers, O&M personnel and others related to the program within the organisation. Workshops, presentations and interviews with corporate management, accounting and finance department, information technology department and others can be used to identify areas where collaboration and flow of information will be necessary.

The *implementation phase* is based on a detailed implementation strategy, which involves two sets of activities including:

- assessment of current position:

    - verification of sources of data, including accuracy and availability;

    - benchmark assessment of all practices, procedures and systems

    - identification of strength and weaknesses based on benchmark assessment

- identification of resources and implementation

    - assemble asset management team;

    - staff training;

    - IT hardware and software;

    - identify external consultants;

    - develop data collection methodologies or each asset group;

    - develop asset replacement/investment criteria.

The asset management program represents a substantial investment for the organisation. The largest cost of implementing the program is incurred in the collection of asset information and development of the asset register. A large amount of information is often held in manuals, drawings and plans by various units within the organisation. Valuable information is often held by experienced staff within the organisation who can often assist in tracing historic data and information not readily accessible through the official records.

6.5.2   Asset Management Information Systems

Asset management information systems must be able to process and provide basic information on the characteristics and the performance (financial, structural and hydraulic) of assets. The core of an effective asset management program is the ability to store and retrieve information on specific assets fast and in a format that is appropriate for strategic planning purposes and for day-to-day management.

The management and operation of irrigation and drainage assets requires specific structural, financial and hydraulic information that must be contained in the asset database. It is desirable that the asset database be integrated with digital mapping systems, GIS and hydraulic operation algorithms to assist operators with the visualisation of the spatial location of assets and hydraulic functions.

The large amount of data storage and manipulation demand the use of computerised management information systems. Figure 6.9 provides a general example of integration of an asset management system, the maintenance system and system operation software for a typical canal irrigation system. Integration of the asset management information system with the information technology system across the organisation is essential to maximise efficiency. This can only be attained if proper integrated planning is carried out before developing the various modules of the information management system.

The asset condition tracking, maintenance and work management modules are the main elements of the asset management information system. Asset condition and maintenance history are continuously updated from either visual inspections or maintenance work carried out on individual assets. The work management module processes all orders for maintenance and repair work issued to the relevant section of the managing organisation. Work analysis and management keeps track of the backlog of orders and priority of maintenance work. Data on resource information including staff time, utilisation of spare parts and equipment can also be stored in the database to enable evaluation of maintenance performance and estimation of future maintenance needs. This may also include inventory of spare parts and expenses associated with each work order. A seamless integration of both modules is desirable to ensure automatic update of the asset condition.

If an Information Technology System is in place across the organisation, it is important that the asset management modules becomes fully integrated. This is particularly desirable where the organisation must comply with asset accounting standards in regard to valuation and depreciation. Likewise, the normal financial planning of the organisation requires updated forward estimates of investment and recurrent expenditure information in order to anticipate future financial outlays and procure adequate financing.

The operation module contains the canal operation and monitoring decision support utilities that are used for planning and supporting the day-to-day operation of the water supply system (Malano, 1994, and Turral *et al*, 1997). The physical description of the infrastructure and agronomic and climatic data needed for the computer model(s) used to operate the water delivery integrated database also resides on the computer as part of the integrated database. Digital mapping and GIS tools greatly enhance the capability by providing an interface for graphical data entry, editing and viewing asset positioning.

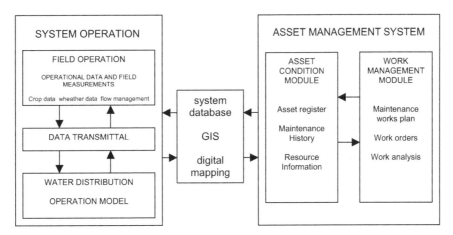

Figure 6.9 Integrated information technology structure for asset management and operation

Chapter 7

Performance of Irrigation and Drainage Service

> *Once the strategic business modelling phase is complete, it is time to check those visions against reality. Visions without a reality base are hallucinations. The performance audit is intended to prevent hallucinations.*
>
> Goodstein et al, 1993

Performance assessment is the measurement of the degree of achievement of the strategic and operational objectives set out in the strategic planning. Performance assessment is concerned with determining the allocative and productive efficiency for achieving results. It must be viewed as an essential tool to measure the quality of the organisation's management process to carry out its functions and conduct its business. While the essential function of irrigation and drainage organisations is the provision of water supply and water removal services, they must do so within a wider environment in which water and other resources are shared with other economic activities. Therefore, the performance of irrigation and drainage organisations will for certain purposes be also evaluated by external organisations and society in general with reference to objectives set out for the totality of the water resource system and the productivity of water use by agriculture.

7.1 CONCEPTUAL FRAMEWORK

The concept of performance emerged initially from the industrial world where performance oriented processes are aimed at accomplishing the process functions using less resources and time. Production systems in general utilise resources as inputs that are transformed into outputs in the form of goods and services needed by people by means of production processes. The efficiency with which these processes are carried out is termed *process performance*. Defined in this way, performance is predominantly a physically based concept.

In business, where profit is the overriding objective, performance will not only include the efficiency of resource use in the production process, but will also consider the marketability of the goods or services produced. The extent to which goods meet customer satisfaction is therefore another important performance parameter. An all-encompassing measurement of performance in industry and business based activities is relatively simple and can be integrated into a single indicator that reflects these two aspects: *the return on investment* (Murray Rust and Snellen, 1993). It is possible to use this single indicator because it expresses performance in monetary values taking into account the efficiency of the production process and the marketability of the product or service as an expression of customers' satisfaction. It is the most common indicator used to measure the performance of private or public corporations that allows shareholders and other interested parties to draw conclusions about the organisation's performance.

However, irrigation and drainage organisations carry out activities that are a great deal different from those that form the "main business" of commercial and industrial organisations, in that:

♦ irrigation and drainage organisations provide services to customers in a monopoly environment;.

♦ irrigation and drainage involve complex and interacting physical, socio-economic and environmental processes that are of a location-specific nature;

♦ the delineation of the irrigation system in which the concerned processes take place varies depending on the purpose of the performance evaluation process.

The last distinction is particularly important given that different frameworks can be used to define the extent and purpose of the system whose performance is being measured depending on where the boundaries of the system are set.

The economics concept of performance is far more encompassing and views the allocation of resources not only from the point of view of physical production or business concept but also the wider perspective of alternative uses of the same resource. It refers to the resources being *allocated efficiently* if the same resources could not have been used to produce a different result that is valued more by society. There are two elements built into this concept: (a) the actual efficiency of the process concerned, and (b) the value that society assigns to the final output. For instance, an irrigation system may produce crops at very high yields for which society has no need. The resources allocated to the production of these crops are not *used effectively* despite high crop yields.

In irrigated agriculture, resources are utilised in the form of land and water, investments in irrigation infrastructure and farm equipment and operational inputs such as personnel, plant and equipment. These resources are used in two distinct processes, namely:

♦ the production of food and fibre through the agricultural system;

♦ the provision of irrigation water supply service and drainage removal service.

The performance of these two primary processes is also amenable to be assessed in terms of the productivity of the agricultural system, and the quality and cost of the services provided by the irrigation and drainage organisation, respectively. Different stakeholders in the irrigation and drainage activity have different interests in these two processes. Production of food and fibre and productivity of irrigated agriculture is a main interest of the farmers and the government. In this context, water-use is also of more interest to those who must judge the productivity of water for irrigation in relation to other uses of water. These include policy makers, planners or river basin authorities. Conversely, the quality of service provision is of primary interest to the irrigation and drainage authority and their clients and in certain cases the Government.

The results generated by a particular irrigation or drainage system in relation to the resources utilised in the process is often termed *irrigation performance* (IIMI-ILRI-IHE,

1999). This concept is also encapsulated in the following definition of irrigation performance: "*...is the results delivered by an irrigation system toward a set of objectives including productivity, equity, reliability, sustainability, profitability and quality of life*" (IIMI, 1989). This definition is intended to evaluate results of objectives which operate at different sectoral levels and which impinge on different stakeholders involved in or affected by the irrigation system. These include:

+ the farm production process which determines productivity, profitability and to some extend the environmental sustainability of the irrigation and drainage system;

+ the irrigation and drainage organisation that provides irrigation services, and;

+ the members of society that are affected by the irrigation and drainage system.

It must also be noted that the definition refers only to performance of irrigation systems; although it could equally be applied to the performance of drainage systems.

The main focus of performance assessment in this monograph is the performance of irrigation and system management given that this monograph is directed to the development of a framework for the provision of irrigation and drainage services. Nevertheless, it is important to observe that this aspect of performance assessment must be viewed only as a subset of a wider performance assessment framework for irrigation and drainage. The following section provides an overview of this comprehensive framework and the following sections focus primarily on the assessment of organisational performance for service delivery.

7.2 PERFORMANCE SPHERES

Despite the relatively large volume of work carried out by various researchers (IIMI, 1989; Murray-Rust and Snellen, 1993; Small and Svendsen, 1990; Bos *et al* , 1994) there is no commonly agreed framework for performance evaluation of irrigation systems. In a more recent effort to establish such a framework, IIMI-ILRI-IHE (1999) have attempted to come to terms with this problem. The study presents a framework for achieving performance orientation in the management of irrigation systems. This section draws on some of the concepts presented in this study.

The proposed framework for evaluation of irrigation and drainage performance is presented in Figure 7.1. In this framework performance assessment is concerned with two main domains:

+ the efficiency of the management process within the organisation that determines its performance in delivering irrigation and drainage services;

+ the allocative and productive efficiency of resources employed in irrigated agriculture in general.

In this context, the resources employed in irrigated agriculture are all the resources employed in the agriculture production process and in the organisational management process directed to the provision of the irrigation and drainage services. These include soil and water, financial, infrastructure and human resources, and agricultural inputs. The efficiency with which these processes are carried out determines the overall productivity of irrigated agriculture and also of the irrigation and drainage organisation.

The main focus of the next sections is the evaluation of all the management processes of the organisation that enable the provision of irrigation and drainage services.

7.2.1 Organisational Performance

The effectiveness of the managerial and operational processes within an organisation determines its performance in delivering irrigation and drainage services. Murray-Rust and Snellen (1993) defined *organisational performance* on the basis of two criteria which include *the degree to which the services offered by the main system managers respond to farmers' needs; and the efficiency with which the irrigation system uses resources in providing these services.* This focuses on two distinct aspects of the organisation's management process:

- *service delivery performance* or the ability of the system operation process to deliver the level of service declared by the organisation or agreed between the organisation and customers;

- *productive efficiency* or the amount of resources or cost involved in delivering this service.

Service Delivery Performance

The quality of services provided by the organisation is determined by how well the actual service delivery matches that level of service specifications as agreed with the customers or declared by the organisation. The ability of the organisation to meet level of service commitments is largely a function of the:

- adequacy of the hydraulic infrastructure in relation to the level of service to be provided;

- adequacy of the operational rules and procedures designed to meet the level of service specifications (Lee *et al*, 1998);

- adequacy of operational inputs - staff number & skills – to implement these rules and procedures.

As discussed in Chapter 4, level of service specifications become a set of targets (norms) against which the performance of service must be evaluated. Level of service specifications must be developed from extensive consultations between customers and the service provider organisation and are therefore specific to each system and type of service provided – irrigation or drainage. This implies that indicators of service delivery performance are also system-specific, and are designed to measure the ability of the organisation to meet its service commitments and track this performance over time.

Productive Efficiency

The types of services provided by the irrigation and drainage organisation may be unilaterally declared by the organisation or arise from an extensive consultative process between the organisation and its customers. These are defined in terms of service specifications that must form part of formal contracts or customer service agreements between the authority and their customers. The performance with which the irrigation and drainage authority delivers its services must be evaluated against these service specifications. The nature and definition of service specifications is the subject of Chapter 3.

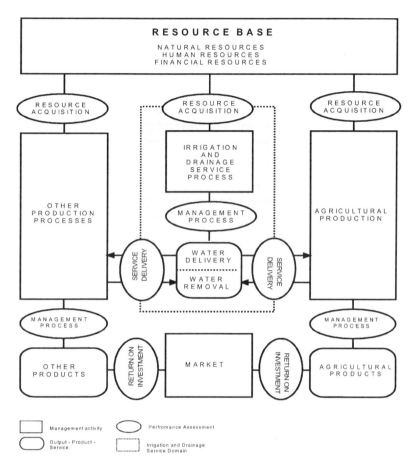

Figure 7.1   Framework for Performance Assessment of Irrigation and Drainage Services

Service orientation implies that a given level of service must be provided in the most cost-effective manner. The ability to deliver irrigation and drainage services at the lowest possible cost is referred to as *productive efficiency*. Cost-effectiveness is therefore a key element for judging the productive efficiency of the provision of irrigation and drainage services.

In a free market environment, the producers of goods and services generally strive for the maximisation of their profit. This goal and perfect competition ensure that goods and services are produced at the lowest cost achieving high productive efficiency. Moreover, if the production of goods and services occurs in a free market economy it can be assumed that *allocative efficiency* is also attained at a macro-economic level.

Irrigation and drainage organisations also use resources as operational inputs to provide services; but there are two important characteristics that set them apart from their commercial and industrial counterparts:

- they are in a monopoly position hence there is often no automatic incentive for achieving high productive efficiency;

- they do not seek to maximise their profit; although it is possible that they are required to yield a predetermined rate of return.

Therefore, clear and effective accountability mechanisms must ensure that service is provided in the most *cost-effective* manner. The efficiency with which the organisation uses resources for the provision of services determines the cost of service provision. This cost is an important performance indicator of how the irrigation and drainage organisation utilises its personnel, infrastructure assets and other inputs to deliver its services.

Management Performance

The shift in focus towards service orientation implies a comprehensive "culture" shift in the manner in which the management processes of the organisation are conducted. This involves both the internal management processes and the outward accountability of the organisation.

Performance assessment of organisations is an essential element in the determination of the degree of achievement of the objectives set out in the strategic and operational planning stages. Management performance is concerned with the ability of the organisation to formulate and implement plans for delivery of irrigation and drainage services. It deals with the quality of the management conditions established, or available, and the management processes that occur continually in the organisation as it carries out its functions and conduct its business to serve its essential purpose. The performance of the organisation is a critical determinant (or causal factor) of other areas of performance of the irrigation system as a whole.

There are a number of performance areas within an organisation that are defined by a set of related skills, procedures and capabilities (Constable and Malano, 1997). These areas relate to the organisation's structure and processes that enable an efficient production process and a high degree of responsiveness towards the desires of, and accountability to its customers. A performance area describes a generalisation or pattern of performance that can be observed or verified through evaluation. Some of these performance areas are not amenable to be evaluated in quantitative terms and thus only qualitative performance assessments can be made. Areas of performance inquiry include:

- *management and administration:* Management is organising people and resources to accomplish the work of the institution. Effective management is demonstrated by the capacity to get the most out of the resources available (human and other) in a deliberate or planned manner. Good managers have a clear sense of goals and priorities; they know who to rely on to get a job done and how to delegate to them the means to do it. They are aware of operational details; they monitor the work and follow-up consistently. An effective management climate is characterised by teamwork, co-operation and good communication among the staff. The counterpart to management skills is the existence and use of key administrative systems. These are the policies and procedures, which regulate and guide the actions of management. A mature organisation has designed or evolved effective sub-systems such as personnel, budget, accounting, financial management, and management information systems;

- *commercial orientation:* It is the degree to which actions in an institution are driven by cost-effectiveness and operating efficiency. This orientation can be viewed at both policy and operational levels, and both levels are important. Commercial orientation is important at policy level, even if significant revenues are routinely derived through subsidies. Subsidies if any, should be identified and tied to specific areas for which the controlling authority has taken a political decision to subsidise, rather than provide a blanket subsidy. Operationally, everyday activities are guided by quality standards and by constant attention to cost factors and achieving the best use of financial resources;

- *customer orientation:* It represents the central focus of service-oriented management and entails organising and directing the services of the institutions towards the customers or the end-users. Staff of effective organisations see serving customers as their primary function. With this premise, all work, programs and innovations are directed towards greater efficiency, effectiveness and equity in service to the consumer. Staff at every level is aware of this customer orientation and sees it as governing positively their important daily operational decisions and actions. Effective institutions have workable means for interaction with their customers. These may include emergency outlets or 'hot lines' when there are crises, clearly identified places where disputes about bills or services can be arbitrated and ways that interested customers can make suggestions to influence overall policy;

- *technical capability:* It is a measure of the institution's competence in conducting the technical work required to carry out the responsibilities of the organisation. Most of this technical work is performed directly by skilled, qualified employees, but outside specialists including consultants and contractors may be used where appropriate;

- *human resources:* Developing and maintaining staff include those activities toward recruiting staff, providing skills to do the jobs and grow professionally, and providing adequate job satisfaction and wages and benefits to retain competent personnel. Effective institutions develop and maintain their personnel. This includes both formal training programs and information training that occurs through on-the-job training, apprenticeships and job rotation. In addition to a regular process of skill transfer, effective institutions maintain staff through providing sufficient incentives, compensation, employee benefits, and promotion opportunities so that there is a minimum of unwanted turnover. Institutions that develop and maintain staff consider that people are their most important assets;

- *organisational culture:* The culture of the organisation is represented by a set of values and norms, which inform and guide everyday actions. The culture forms a pattern of shared beliefs and assumptions that translate into behaviour which can be observed. An organisation's culture is conveyed in a number of intended and unintended ways. Although seldom stated, cultural beliefs, behaviours, and assumptions serve as a powerful means for defining and justifying organisational culture either in positive or negative ways. An important factor in organisational culture is how the organisation deals with change or crisis. People are often required to alter the way they operate when a major change is introduced such as new technology, new leadership or organisational restructuring. How the organisation responds to such changes either by realigning its forces to support innovation or resist change is an indication or the overall health of the organisation.

- *interactions with key external institutions:* The organisation's capacity to influence positively and strategically those institutions that affect its financial, political and legal ability to perform is a measure of how well the institution interacts with external institutions. The performance of an organisation is affected by many entities in the external environment including the political (parent ministry), financial (budget and finance ministry) and regulatory (state/provincial government) agencies. This task requires the capacity to anticipate activities, which might affect the organisation and establish strategies to deal with them.

7.2.2   Allocative Performance

In previous sections it was established that irrigation and drainage organisations attain productive efficiency when services can be provided in the most cost-effective manner. Productive efficiency is therefore primarily determined by the quality of the internal management processes of the organisation.

An alternative assessment of the use of resources by irrigation and drainage systems can be made by considering the value that society places on these resources. From the point of view of society, *efficient allocation* of resources implies that the same resources could not have been used to produce any alternative outcome that is valued more by society (Lipsey *et al*, 1987). The allocative efficiency concept applies both to the inputs into the agricultural production process, and the use of resources in the delivery of irrigation and drainage services. The resources used include natural resources - land and water – investments in infrastructure for irrigation and drainage systems, and on farm and operational inputs including personnel and farm inputs.

The allocative efficiency of water within the irrigated agriculture subsector is influenced by decision making and performance at three levels: (a) the government and river basin authority, (b) the irrigation and drainage organisation, and (c) the users.

Of increasing importance to society is the allocative efficiency of water among different uses. This assessment must be carried out in the context of the entire river basin. Illustrations of this concept can be found where for example water is used for agriculture when it could be allocated to urban use to improve the quality of life of a larger number of people. Another example is when alternative industrial use could produce a greater economic output for society. The assessment of the allocative efficiency involves an evaluation of the results of these allocative decisions. These types of decisions are made at different levels of administration of the water resources system. Governments and relevant river basin authorities would be involved in water allocation between water sub-sectors in a river basin.

The policies and actions of the irrigation and drainage authority play an important role in determining the allocative efficiency of water. For instance, the efficiency of water use on the farm and the productivity of agriculture are affected to a large degree by the quality of irrigation and drainage services. Service pricing policy can have an effect on the type of crops grown under irrigation. Full recovery of irrigation and drainage service costs would translate into higher water prices, which in turn would promote the adoption of higher value crops.

Irrigators play a decisive role in the productivity of their enterprises through the quality of their water management and husbandry practices. More productive irrigators would obtain a higher productivity and return per unit of resource utilised.

Society also places an increasing value on maintaining the environmental quality of water resources. In Chapter 2 we discussed the use of water for irrigation in the wider context of integrated water resources management. In this context, irrigation and drainage activities interact and often compete directly with other water uses by consuming resource   or indirectly as a polluter. The negative impacts of agricultural effluent on the quality of water are one of the important management factors that must be addressed in order to meet societal expectations about environmental quality.

The *environmental performance* of irrigation can hence be viewed as the ability to increase agricultural productivity while minimising environmental impacts. Low levels of performance in the allocation and use of water and other inputs can also lead to undesirable environmental impacts. Thus low environmental performance can be the result of low allocative performance. For instance, poor water management practices in the form of excessive water application induce waterlogging and subsequent salinisation of the cropland. Excessive use of fertilisers can cause excessive nutrient runoff and deterioration of the water quality in rivers, streams and reservoirs.

7.3 PERFORMANCE ASSESSMENT

7.3.1   Measuring Performance

All the performance categories defined earlier are amenable to measurement by use of proper indicators or indices designed to gauge the degree of attainment of goals, objectives and targets. There are several stakeholders that are interested in different aspects of irrigation and drainage performance. At one level, Government, river basin authorities, agricultural organisations would be primarily interested in the allocative performance of irrigation and drainage. At a different level, Government, the organisation's management and customers are chiefly interested in measuring the achievement of the goals, objectives and targets set out in the service agreements and the strategic plan of the organisation. This requires monitoring of progress towards achieving the goals, objectives and targets set out in the strategic plan.

Monitoring is information gathering and transmittal to assess performance. Monitoring follows outputs and enables management to make timely adjustments in inputs in order to stay "on track". This entails monitoring of the critical indicators that measure the achievement of strategic goals. An important difference must be established at this point between the notion of data and information. Data must be processed into a form that becomes useful to provide management with useful information to support decision making. This is accomplished through the evaluation of the primary monitoring data.

Evaluation is information processing to examine goals and strategies, to compare the results of monitoring and observations with the preset standards or targets and asks whether the strategic plan is "on track." It takes into account the accuracy of planning assumptions and the effectiveness of planning strategies. Evaluation should be viewed as a continuous process of internal analysis and adjustment based on effective feedback. Unlike a construction project where there exists a relatively fixed relation between inputs and outputs, in systems management most likely the relations are themselves unknown. Thus the evaluation process becomes itself a critical component of the management learning process.

The design and selection of appropriate measures of performance is inextricably linked to the nature and purpose of the performance area that is being assessed. *Performance indicators* are a set of specific measurable parameters of behaviour or procedures related to a performance category which, when analysed together indicate the degree to which competence standards are met in the performance category. The performance standards need to be determined on the basis of realistic goals set during the management planning process, for which observations of successful performance of currently operating institutions may be useful. Often, standards and associated indicators are imposed from outside the organisation by government or other parties and may become part of the service agreement with customers.

Small and Svendsen (1990) distinguish three types of performance indicators:

- *process indicators*, relate to the internal operation of a system that lead to outputs, e.g. technical capability of the organisation, performance of infrastructure assets, budgetary allocation, etc.;

- *output indicators* that describe the quality and quantity of the outputs from the system. Indicators used to characterise service specifications such as amount and time of water delivery, pressure of water distribution, etc. are typical output indicators used to measure the performance of service delivery in the context of the service agreement;

- *impact indicators* that relate to the impact of outputs on the wider environment, e.g., crop yields, farmers' income, waterlogging and salinity levels, etc.

The application of these three types of indicators to the ORMVAM de la Moulouya in Morocco is illustrated in Box 7.1.

Bos *et al* (1994) presented a performance analysis framework accompanied by a set of standard indicators. These were identified and selected with a view to providing a basis for assessment of irrigation system performance and for performance comparison between systems. We advocate that rather than adhering to a set of standard performance indicators, these must be tailored and selected to specifically describe the goals and targets that are identified for the area of performance being measured. There are a number of desirable attributes that must be considered when selecting performance indicators such as:

- they must be unambiguously defined for the situation they describe;

- they should be accurately measurable under the existing conditions at acceptable cost;

- they should indicate the "state of condition" in a specific and precise manner;

- they should provide unbiased measures of the parameter concerned;

- they should be agreed to and measurable by both the organisation providing the service and by the service recipients.

Box 7.1 Examples of Performance Indicators as used in ORMVAM, Morocco. (van Hofwegen *et.al.* 1996).

Indicator	*Formula*	*Value 1995/96*
Ouput indicators		
Adequacy	$\dfrac{\text{Volume of water delivered} + \text{Eff Rainfall}}{\text{Volume of water required}}$	0.78
Reliability	$\dfrac{\text{Actual delivery}}{\text{Intended delivery}}$	1.01
Process indicators		
Cost of service (incl. amortisation infrastructure)	$\dfrac{\text{Total cost of service}}{\text{Volume of water billed}}$	0.035 US$/m3
Water delivery efficiency	$\dfrac{\text{Volume water sales}}{\text{Volume released from reservoir}}$	0.82
Pumping stations (13 pc.)	Energy consumption per m^3 water pumped per m^1 lift,	$0.0022 - 0.0061$ kWh/m^3/m
	Cost per m^3 water pumped	$0.008 - 0.043$ US/m^3
Impact indicators		
Output per cropped area	$\dfrac{\text{Gross value of production}}{\text{Irrigated cropped area}}$	2520 US$/ha
Output per unit irrigation supply	$\dfrac{\text{Gross value of production}}{\text{Diverted irrigation supply}}$	0.41 US$/m^3$

7.3.2 Monitoring and Evaluation Systems

Meaningful performance assessment requires that an efficient and effective Monitoring and Evaluation System (MES) be implemented. The implementation of an MES for irrigation and drainage systems implies an increased capability to gather and process information, and then make informed decisions based on that information. It is critical that a MES is designed to be responsive to management needs and capabilities, to direct information to that level of management structures where it is needed and may best be utilised.

There are two main time frames involved in the operation of the MES. The MES permits management to know periodically the performance status of the strategic plan. The type of process monitored and the nature of the indicators used depend on the specific needs of each irrigation system or project. The type of data and information needed for this purpose is periodically compiled, analysed and used by management. System operation and monitoring requires data and information transmitted to field operations personnel and their feedback received in real-time to make appropriate flow control decisions to meet the agreed standard of service.

The details for establishing an MES for irrigation systems vary considerably between agencies according to their specific administrative and organisational structure. In any case, the system should be able to support management through systematic data gathering, assessment, reports and refinement (Feedback). The following general process is suggested for establishing an MES:

- define the MES goals and objectives, which are generic to the agency and those, which are specific to strategic plan within the organisation;

- establish procedures, formats and information channels appropriate to the needs of the organisations' management and to the specific strategic plan;

- train and assign staff personnel for specific tasks in MES;

- implement simultaneously in various levels of the irrigation and drainage agency, District, Regional and Central; and establish clear functional responsibilities for each level;

- establish and regularise periodic reporting schedule (weekly, monthly, mid- season, seasonal, annual) and channels for delivery and feedback;

- establish appropriate methods for conducting evaluations of the information collected during the monitoring process by defining meaningful indicators;

- establish procedures for acting upon the indicators uncovered during the evaluation process by defining clear priorities for adjustments in the system;

- establish annual evaluation of performance to assess the MES program, and change the procedures, formats and channels as needed to meet program objectives;

Some times MESs are set up in two or more units of the agency. In this case, each unit will conduct monitoring and evaluation of the internal activities in the unit. This arrangement may be necessary when monitoring and evaluation tasks are related to a specific project or program rather than the agency as a whole. A problem often encountered under this arrangement is that the information does not flow across the various sections of the department.

A superior arrangement often is to establish a "stand alone" unit which gathers data from the various sections of the organisation. If organisations have various levels e.g. Central, Regional, District, this unit must be replicated at these various organisational levels, although only with the roles relevant to the activities of the organisation at each level. The process starts at the district or system level and logically aggregates the data up to the Central level. At each level of aggregation the information has particular significance. In this way, managers at each level can answer for their performance to the central management and at the same time make adjustments to the implementation of their programs at lower levels.

Chapter 8

The Way Forward

I never think of the future. It comes soon enough.
Albert Einstein (1879-1955)

Irrigation and drainage is at a crossroad and facing multiple challenges. Required increase in food production will have to be achieved through increased productivity in an environment of increasing competition for water and land and more stringent environmental regulations. This entails a leap in performance of irrigation and drainage systems which demands a shift in paradigm towards service orientation. This shift towards service oriented management must form the basis for greater innovation supported by increased research efforts in this sector. The future investment needed to upgrade irrigation and drainage infrastructure provides the best opportunity to foster this shift in paradigm. In this context, investment in building the institutional and management capacity of irrigation and drainage organisation must be assigned the highest priority. Also, the adaptation and use of management quality methods and standards as used in other commercial and service sectors, can play an important role in achieving this paradigm shift.

8.1 INTRODUCTION

In this last chapter we want to highlight and reiterate some of the key future challenges the irrigation and drainage sector is facing and how the concepts presented in this monograph can play an important role in meeting these challenges. Our main goal in expounding this service approach to management of irrigation and drainage is to make the supply of irrigation water and the removal of excess water more responsive to the needs of "customers". The term customer has been used consistently throughout this monograph to emphasise their role as service recipients; whereas the term organisation has been used to indicate its role as service provider. These two terms encapsulate the essence of the management approach presented in this monograph, that is: management of irrigation and drainage systems with a service orientation.

With this perspective in mind, we also want to emphasise the opportunities that will be available in future to policy makers, planners and managers to apply and implement this management approach; together with other technological innovations. Capturing these opportunities will be critical for meeting the challenges faced by irrigation and drainage.

8.2   MANY CHALLENGES – ONE COMMON RESPONSE

8.2.1   Poverty reduction and food supply

Poverty reduces the ability of people to access food and other basic necessities of life. It is widely recognised that agricultural growth is needed to raise the farm and non-farm income of rural populations. Despite urbanisation, over 70% of the poor live in rural areas (World Bank, 1997). As agriculture remains the main source of income for vast numbers of people, it is still an effective means of raising living standard of rural population.

Estimates of food supply and demand by the year 2020/30 indicate that regional shortages of food may occur in the African and Asian continents despite global food production being sufficient to feed the world population. Access to food in these regions may be restricted by lack of foreign exchange needed to meet the expected shortfall.

Irrigation and drainage is expected to play an important role in fulfilling poverty alleviation targets. It is conceivable that a large proportion of this additional food supply will come from existing irrigation developments. This would entail a huge improvement in the current level of performance of irrigation and drainage systems which can only occur if current management standards are improved.

8.2.2   Resource use

In a global economic context, irrigated agriculture will be faced with increasing competition for water, land and other resources. Increasing competition for water between subsectors is likely to reduce the water availability for irrigation over the next decades. In a comprehensive study of water availability worldwide, Seckler (1998) classified countries into 5 groups according to their projected water availability and withdrawals. The study indicates that 8% of the world population living in group 1 countries will face severe water shortage by 2025. The same study concludes that 50% of the increased food demand can be met by increasing overall irrigation efficiency to 70%. Indeed, a tall order by any stretch of the imagination if past experience with efforts to improve efficiency of water use is of any value.

These projections are based on average country wide water balances. A similar study at the basin level is likely to worsen the picture as withdrawals from many river basins can be limited by several factors, including:

♦ the temporal variation of flow, especially that low flow periods often coincide with maximum irrigation requirement;

♦ water quality constraints arising either from excess pollution or from salt water intrusion in low lying river deltas;

♦ increased demand for in-stream requirements to maintain the aquatic environment and prevent salt water intrusion;

♦ Degradation of watersheds leading to increased sediment loads and greater variability of river flows.

8.2.3 Environmental impact and sustainability

Irrigation and drainage practices are often associated with negative impacts on the soil and water. These impacts occur in several ways. There are extensive irrigated areas that are affected by waterlogging and salinisation as a result of poor water management practices often in combination with unfavourable geological conditions. At present, waterlogging and salinisation is estimated to affect 15% of the irrigated area worldwide. While part of this area can be reclaimed by providing appropriate drainage, the disposal of drainage effluents often degrades the quality of the receiving waters rendering them less usable for human consumption, wildlife or agriculture.

As society awareness of environmental issues increases, it is expected that irrigation managers will need to be more accountable not only to irrigation customers but also to the wider community. Managers will be under increasing scrutiny for efficient and effective resource use and the effects that the irrigation and drainage activities have on the surrounding environment. As a result, agencies will have to improve the sustainability of irrigation and drainage operating in an environment of integrated water resources management. In such an environment the agencies will be directly accountable to their clients and to society at large through the appropriate river basin organisations which will integrate and manage multiple water uses and users. Increasingly, regulations and standards for agricultural effluent will be imposed on the irrigation and drainage sectors by river basin and environmental protection authorities. In the future, irrigation and drainage agencies are also more likely to be required to meet water quality standards as part of their service specifications.

The sustainability of irrigated agriculture recognises several dimensions, including:

♦ environment: acceptable environmental impacts;

♦ infrastructure: long-term performance of assets;

♦ production: real increases in goods an services involved;

♦ social: acceptable to all stakeholders;

♦ finance: maintain financial viability of farms and other users.

The negative impacts of irrigation and drainage and the consequent lack of sustainability are largely a direct result of poor irrigation and drainage system performance. The often poor water management practices result from low quality of irrigation and drainage services. The quality of services with its infrastructural, social and financial aspects are primarily controlled by the enabling environment in which these agencies operate, the quality of the internal management processes of the organisation and its customer relations. Achieving performance improvement in an increasingly competitive water environment requires well formulated management processes within the organisation providing irrigation and drainage services. It is the responsibility of the Government to create an environment in which these agencies can develop their internal management processes focussed provision of services to satisfy their customers and the expectations of society at large.

In summary, achieving the goals of poverty alleviation, increased food security, efficient resource use, reduced negative environmental impact and improved sustainability involves a common response. This response is embodied in achieving increased performance of irrigation through service oriented management based on transparent management processes and efficient and effective accountability mechanisms within an integrated water resources management context.

8.3   MODERNISATION, INNOVATION AND RESEARCH

As discussed in Chapter 7, the performance of irrigation can be viewed from two perspectives:

- ♦ the efficiency of the management process within the organisation in charge of providing irrigation and drainage services; and,

- ♦ the allocative and productive efficiency of resources employed in irrigated agriculture.

Whilst both aspects of performance are important, the efficient delivery of irrigation and drainage services is recognised as a key factor in improving the productivity of irrigated agriculture and allocative efficiency of resources used in both the delivery of services and the agricultural production process.

The slow pace of innovation and change in irrigation and drainage is well documented. Adoption of technology and innovation does not only lag behind other sectors of the economy but more importantly it lags behind technological advances in production agriculture, a sector that irrigation and drainage directly supports (World Bank/UNDP, 1990). In service oriented management the responsible agencies are required to respond to the ever increasing demands of their clients for reliable and flexible services. Innovation and modernisation will become a necessity for these agencies to meet these demands.

Achieving this level of organisational performance and productive efficiency would require the application of the vast store of technology that is readily available to enhance water management on-farm and in the distribution system. Additionally, it will also require a substantial increase in investment in research and development to make this technology suitable for the site specific conditions irrigation and drainage organisations have to operate.

A hardware technologically driven approach is likely to fall short of the attempted aims if it is not accompanied by a fundamental shift in the prevailing approach to management of irrigation and drainage systems. Such a shift must be towards management systems with a well-defined service orientation. This is a critical element to create an environment that encourages technology innovation both at the farm level and by the managing organisation. Much of the technology available for on-farm water management has been developed and used in developed countries. These countries have in general one thing in common: the provision of a higher level of service by the irrigation management agencies. These agencies are accountable and are able to respond to their customers demands.

As discussed in Chapter 5, modernisation of the water delivery system does not imply "high tech" solutions but rather changes that enable service improvement and controllability of the hydraulic system to provide a well defined level of service. The guiding principle for modernisation of existing irrigation systems must be the provision of an agreed or declared level of service that must be achieved at an agreed level of cost with customers. This is the essence of the formulation of level of service.

8.4   OPPORTUNITIES

Meeting the challenges faced by irrigation and drainage will require a shift in paradigm towards service oriented management of irrigation and drainage systems. This move can be made by capitalising on future interventions required in the irrigation and drainage sectors in many of the most significant irrigation countries. It is estimated that 50-70% of the existing infrastructure will require some kind of infrastructure interventions in future to upgrade (rehabilitate/modernise) the current generation of systems. This constitutes the greatest opportunity available to introduce the required technological and management changes

Modernisation must aim to the provision of an improved level of service involving primarily improved flexibility and reliability of irrigation supply. As such it must include both physical as well as institutional and management changes. The lack of emphasis on any of these aspects will not allow the potential improvements in productivity to be achieved. Furthermore, current policy efforts to devolve institutional/management responsibilities to farmers with the objective of lightening financial burdens on government treasuries are not likely to translate into performance gains that are required to meet the future sector challenges unless they are accompanied by a shift towards service orientation by the irrigation and drainage organisations.

Building the capacity of irrigation and water resources management agencies must be seen as a priority investment to achieve these performance goals. To reverse the trend of past failures in irrigation and drainage investment, our view is that investment in water resources infrastructure should be preceded or paralleled by development of the required institutional management capacity. Lending agencies can play a pivotal role in the implementation of this paradigm shift. Institutional capacity aspects related to development projects need to be given a much greater profile than in the past. Often, this will mean to stay with the project longer and perhaps institutional aspects of projects will need to be in place before any infrastructure intervention begins.

8.5   ENSURING FUTURE MANAGEMENT QUALITY

Higher performance of irrigated agriculture is critically linked with the performance of irrigation and drainage service delivery. Achieving this requires the introduction of service oriented irrigation and drainage system management with a strong focus on performance oriented management, transparency and accountability. These changes can only be brought about if there is a high degree of awareness of the problems and the possible solutions, and the necessary political will. This not only calls for development of management and institutional capacity in the irrigation and drainage subsector but also in the wider water resources sector.

Despite the great strives made in the management of commercial companies in other sectors of the economy and in some cases, the transformation of public agencies towards customer orientation, this movement has not benefited the management of irrigation agencies either in the public sectors or many of those that have been corporatised or privatised.

The implementation of service oriented management involves a shift in the organisation's ethos which can only be achieved if the goals of service provision are clearly canvassed in the strategic plan of the organisation together with the strategies for its implementation.

Management tools that have been available to the commercial and industrial sector can play an important role in upgrading the quality of management in irrigation and drainage organisation. Management quality assurance standards have been developed by the International Organisation for Standardisation (ISO) and the ISO 9000 series are widely applied in the commercial and industrial world. The need for and implications from the implementation of these management standards require a great deal of more thought to identify the potential benefits of its adoption and the needs for adaptation required for the irrigation and drainage sector. Further concern about environmental impacts from irrigation and drainage effluents is also likely to foster the implementation of effective environmental management systems. ISO Standard 14000 series of international standards is designed to be implemented in any type of organisation in either public or private sector companies, administrations and public utilities.

References

Ackoff R., 1981. *Creating the Corporate Future*. John Wiley & Sons. 297 pages.

Ankum P., 1991. *Flow Control in Irrigation and Drainage*. Lectures Notes. International Institute for Infrastructural, Hydraulic and Environmental Engineering. Delft, The Netherlands. 293 pages.

Ayers R. S., D. Westcot, 1976. *Water Quality for Agriculture*. FAO Irrigation and Drainage Paper 29. Rome, Italy. 97 pages.

Banyard J., J. Bostock, 1998. *Asset Management – Investment Planning for Utilities*. Proc. Instn. Civ Angrs. Civil Engineering 1998. Paper 11423. Vol 126:65-72.

Bos, M.G., D. H.Murray Rust, D.J. Merrey, H.G..Johnson, W.B. Snellen, 1994. *Methodologies for assessing performance of irrigation and drainage management*. Irrigation and Drainage Systems Vol.7 no. 4 pp 231-262. Kluwer Dordrecht.

Briscoe 1997. *Managing Water as an Economic Good*. In M. Kay e.a. (eds) Water: Economics, Management and Demand, pp339-361, E&FN Spon, London.

Burton M., R.J.Hall, 1998. *Asset Management: Addressing the issue of Serviceability*. Asset Management Workshop. Afro-Asian Conference, International Commission on Irrigation and Drainage, Denpasar, Bali, Indonesia. 14 pages.

Buyalski C.P., D.G. Ehler, H.T. Falvey, D.C. Rogers, E.A. Serfozo, 1991. *Canal Systems Automation Manual*, Vol.1. US Dept. of the Interior, Bureau of Reclamation.

Carruthers I., 1993. *"Going, Going, Gone! Tropical Agriculture as We Knew It"* Tropical Agriculture Association Newsletter (United Kingdom) 13(3):1-5

Constable D., H.M.Malano, 1997. *Corporate Planning and Human Resources Development*. Lecture notes. International course on Water Resources and Irrigation Management. University of Melbourne. 141 pages.

Daft R.L,1995. *Organization theory and design*. 5th ed .St. Paul: West Pub. Co. 543 p.

Du Buat. P.L.G., 1786, *Principes d'hydraulique, verifies par un grand nombre d'experiences faites par ordre du gouvernement*.2nd edition, Paris 1786.Chapter 3

Food and Agricultural Organisation of the United Nations, 1992: *Wastewater Treatment and Use in Agriculture*. Irrigation and Drainage Series no. 47.

Food and Agricultural Organisation of the United Nations, 1994. *Water Policies and Agriculture*. Special chapter of The State of Food and Agriculture 1993. Rome, Italy. Pages 228-297. FAO, Rome.

Food and Agricultural Organisation of the United Nations, 1996. *Food Production: The Critical Role of Water*. Advanced edition background paper for the World Food Summit September 1996. 61 pages. FAO-Rome

Gaff D. C., 1987. *Value Management – Maximising the value of everything your business should do for every dollar it will spend*. Conference on Engineering Management 1987. The Institution of Engineers, Australia. National Conference Publication No 87/12.

Gerken L. (ed.), 1995. *Competition among institutions*. Houndmills, Basingstoke, Hampshire : Macmillan Press ; New York. 232 pages

Goodstein L, T.Nolan, J W Pfeiffer, 1993. *Applied Strategic Planning: A Comprehensive Guide*. Mc Graw Hill, Inc. New York. 379 pages.

Graaf, M.de, W. van den Toorn, 1995. *Institutional context of irrigation management transfer*. P. 69-68 in: Johnson S. H. et al. Irrigation Management Transfer. Selected Papers from the International Conference on IMT, Wuhan, China.

Grigg N., 1996. *Water Resources Management: Principles, regulations and cases*. Mc-Graw Hill. New York. 540 pages.

Hardin G., 1968 .*The Tragedy of the Commons*; Science Vol 162, 13 dec 1968 pp 1243-1248.

Hofwegen, P.J.M. van. 1996. *Accountability mechanisms and user participation in three agency managed systems in Morocco, Indonesia and the Netherlands*. In ICID, 16th Congress on Irrigation and Drainage, Cairo, Egypt, 1996: Sustainability of Irrigated Agriculture -Transactions, Vol.1.B, Q.46. R.2.03. ICID-New Delhi, India: pp.231-244.

Hofwegen, P.J.M.van, 1997. *Financial Aspects of Water Management, an Overview*. In Hofwegen P.J.M.and E. Schultz , Financial aspects of water management, proceedings of the 3rd Netherlands National ICID Day. A. A. Balkema, Rotterdam, The Netherlands. 113 pages.

Hofwegen P.J.M.van, H. Belguenani, A. E Kassimi 1996. *Use and Utility of Performance Indicators: Triffa Scheme secteur 22, ORMVAM de la Moulouya*, IHE-IIMI-ORMVAM Research Programme on Irrigation Performance. IHE-Delft.

Hofwegen P.J.M.van, H.M.Malano, 1997. *Hydraulic Infrastructure under Decentralised and Privatised Irrigation System Management*. In: Deregulation, Decentralisation and Privatisation in Irrigation, DVWK Bulletin no 20. Pp 188-216, German Association for Water Resources and Land Improvement.

Hofwegen P.J.M. van, F.G.W. Jaspers, 1999. *Analytical Framework for Integrated Water Resources Management: Guidelines for Assessment of Institutional Frameworks*, IHE Monograph 2, Balkema Publishers Rotterdam/Brookfield, 96 pages.

IDB 1997, *Integrated Water Resources Management: Strategy* Background Paper, IDB-Washington D.C.

IIMI 1989,. *Managing Irrigation in the 1990's: A Brief Guide to the Strategy of the International Irrigation Management Institute*. Sri Lanka.

IIMI-ILRI-IHE, 1999, Synthesis Report on Research Program on Irrigation Performance RPIP (In preparation)

ILRI, 1979. *Drainage Principles and Applications*. ILRI Publication no 16.

IMEA, 1994. *National Asset Management Manual*. Institute of Municipal Engineering, Australia .

ICWE 1992: International Conference on Water and the Environment*: Development issues for the 21st Century*. The Dublin Statement and Report of the Conference. Dublin, Ireland. 44 pages.

Jensen M. (Ed.) 1980. *Design of On-Farm Irrigation Systems*. American Society of Agricultural Engineers. 829 pages.

Kotler P., 1994. *Marketing Management: Analysis, Planning, Implementation and Control*. Prentice Hall International. Eight Edition. 801 pages.

Lal R., Pierce (Eds.), 1991 *The vanishing resource*. In Lal and Pierce Soil Management for Sustainability. Soil and water Conservation Soc. Ankeny.pp 1 - 5.

Lal R., Steward (Eds.) 1992. *Need for Land Restoration*. Advances in Soil Science, Springer Verlag, New York, pp 1 - 11

Lal R. *et.al.* 1988. *Are intensive agricultural practices environmentally and ethically sound*? J. Agr. Ethics 1:193-210

Lee P, P.J.M. van Hofwegen, D Constable, 1997. *Financial Management Issues in Irrigation and Drainage*. ICID Journal. Vol 46:1. Pp 49-64.

Lee P., H.M. Malano, E. Caliguri (Eds.), 1998. *Planning and Management, Operation and Maintenance of Irrigation and Drainage Projects*. World Bank Technical Paper 389 .

Lenton R.W., 1986. *Accomplishments, Problems and Nature of Irrigation in International Development*. Symposium on Irrigation, Its Role in International Development-Benefits and Problems. Annual meeting of the Amer. Assn. For the Advancement of Science, Philadelphia.

Lindley E., 1990. *Asset Management Planning: Theory and Practice*. Journal of Water and Environmental Management, Vol 6, 621:627.

Linsley R. K., Franzini J. B., 1979. *Water Resources Engineering*. McGraw-Hill Series in Water Resources and Environmental Engineering. N. York. 716 pages.

Lord W.B., M.Israel, 1996. *A proposed strategy to encourage and facilitate improved water resources management in Latin America and the Carribean*. IDB-Washington D.C.

Lipsey, R. G., P Steiner, D. Purvis, 1987. *Economics*. 8th Edition. Harper & Row Publishers. New York. 942 pages.

Lowdermilk M. K., 1981. *Social and Organisational Aspects of Irrigation Systems*. Lecture for the Diagnostic Analysis Workshop, Water Management Synthesis Project, Colorado State University, Ft Collins, Colorado.

Malano H. M., V. C. Nguyen, T. K. Nguyen, V. C. Dung, M Bryant, H N Turral, 1997. *An Asset Management Framework for the La Khe Irrigation Scheme*. Seminar on Irrigation Water Management in the Red River Delta, Vietnam. 12 pages.

Malaterre P.O., 1995. *Regulation of Irrigation Canals*. Irrigation and Drainage Systems Vol 9. No.4. pp 297-372. Kluwer Dordrecht.

Metcalfe A.V.,1991. *Probabilistic Modelling in the Water Industry*. Journal of Water and Environmental Management, Vol 5, 439:449

Milne L., 1972. *Techniques of Value Analysis in Engineering*. 2nd Edition. McGraw-Hill Book Co. New York. 366 pages.

Ministry of Public Works, 1986, Irrigation Design Standards: Vol KP-05: Tertiary Units, Directorate General of Water Resources, Indonesia

Moorhouse I., 1998. *Asset Management of Irrigation Infrastructure: The approch of Goulburn-Murray Water, Australia*. Asset Management Workshop. Afro-Asian Conference, International Commission on Irrigation and Drainage, Denpasar, Bali, Indonesia. 16 pages.

Morris J., 1987. *Irrigation as a Privileged Solution in African Development*. Development Policy Review Vol 5, pp 99-123.

Mudge A., 1971. *Value Engineering: A systematic approach*. McGraw Hill Co. New York. 286 pages.

Murray-Rust H.,W.B.Snellen, 1993. *Irrigation System Performance Assessment and Diagnosis*. International Irrigation Management Institute. 148 pages.

Newell F.H., 1916. *Irrigation Management*. D. Appleton and Co. New York-London.

Niebel B, 1985. *Engineering Maintenance Management*. Marcel Dekker, Inc. New York and Basel. 327 pages.

Olsen M., 1965. *The Logic of Collective Action: Public Goods and the Theory of Groups*. Cambridge, Harvard University Press.

Ostrom E., 1986. *An Agenda for the Study of Institutions*. Public Choice 48: 3-25.

Ostrom E., 1990. *Governing the Commons: The Evolution of Institutions for Collective Actions*. Cambridge University Press. New York.

Ostrom E., 1993. *Crafting Institutions for Self-Governing Irrigation Systems*; Centre for Self Governance, Institute for Contemporary Studies, San Francisco, California.

Ostrom V., E. Ostrom, 1977. *Public Goods and Public Choices*. In Alternatives for delivering public services, Towards improved performance, ed. E.S. Savas pp. 7 - 49. Boulder: Westview Press.

Pinstrup-Andersen. Per., 1994. *World Food Trends and Future Food Security*. IFPRI Report. 25 pages.

Pinstrup-Andersen P., R.Pandya-Lorch, M.Rosegrant, 1997. *The World Food Situation: Recent Developments, Emerging Issues and Long Term Prospects*. 2020 Vision Food Policy Report, The International Food Policy Research Institute, Washington D.C.

Plusquellec H., 1988, *Improving the Operation of Canal Irrigation Systems*, The Economic Development Institute of the World Bank, Washington D.C. 155 pages.

Population Action International, 1995. *Sustaining Water: an Update*: revised data for the Population Action International Report, Sustaining Water: Population and the Future of Renewable Water Supplies.

Robey D., 1986. *Designing Organisations*. Irwin. Illinois…..

Rogier D., C. Coeuret, J. Bremond, 1987. *Dynamic Regulation on the Canal de Provence*. In: Zimbelman, Planning, Operation, Rehabilitation and Automation of Irrigation Water Delivery Systems. pp.180-200, ASCE.

Rosegrant M.W., C. Ringler, R.V. Gerpacio, 1997. *Water and Land Resources and Global Food Supply*. 23rd International Conference of Agricultural Economists, Sacramento. California.

Schultz E. (ed) 1990, *Guidelines on the Construction of Horizontal Subsurface Drainage Systems*, International Commission on Irrigation and Drainage, Working Group on Drainage Construction.

Seckler D, U Amarasinghe, D Molden, R de Silva, R Barker. 1998. *World Water Demand and Supply 1990-2025: Scenarios and Issues*. Research Report 19. International Water Management Institute. Colombo, Sri Lanka. 40 pages.

Serageldin I., 1995. *Toward Sustainable Management of Water Resources. Directions in Development* The World Bank. 33 pages.

Skogerboe G.V., 1990, *Development of the Irrigation -A Learning Process*, Irrigation and Drainage Systems Vol 4. No.2. pp 151-170. Kluwer Dordrecht.

Small L.E., M. Svendsen, 1990. *A Framework for assessing irrigation performance*. Irrigation and Drainage Systems Vol 4. no 4: pp 283-312.

Snellen W.B.,1996. *Irrigation Scheme Operation and Maintenance,* Irrigation Water Management Training Manual no 10, FAO, Rome.

Snellen W.B.,1997. Service Oriented Management of Irrigation and Drainage Systems. Lectures Notes. IHE. Delft. 7 pages.

Suryadi F.X., 1996. *Soil and Water Management Strategies for Tidal Lowlands in Indonesia*. PhD Thesis. IHE Delft, Balkema Publishers Rotterdam/Brookfield

Tang S.Y., 1994. *Institutions and Collective Action: Self Governance in Irrigation*. Institute for Contempary Studies Press, San Fransisco.

Tardieu H., J. Plantey. 1999. *Balanced and Sustainable Water Management: The Unique Experience of the Regional Development Agencies in Southern France*. ICID Journal. Vol 48:1, pp. 1-5

Thuesen H.G., Fabrycky W. J., 1964. *Engineering Economy*. Prentice Hall, Inc. Englewood Cliffs, N. J. 525.

Turral, H. N., N V Chien, D. V. Dung, N. T. Suu, N. V. Tan, H. M Malano, 1997. *Modelling and Monitoring of System Operation at La Khe Irrigation District, Ha Dong, Vietnam*. 14 pages.

UNCED 1992, United Nations Conference on Environment and Development, Agenda 21.

United Nations Population Division (1996) *World Population Prospects: the 1996 Revision*. New York, The United Nations

Uphoff N., P.Ramamurthy, R.Steiner. 1991. *Managing Irrigation: Analysing and Improving the Performance of Bureaucracies*; Sage Publications, New Delhi. Newbury Park, London.

Verhallen J.M., P.Huisman, L.Korver, 1997. (in Dutch), *Integraal Waterbeheer.* Rijksinstituut voor Integraal Zoetwater beheer en Afvalwaterbehandeling, RIZA, RWS Lelystad. 285 pages

Victoria State Government 1986. *Corporate Planning in Victorian Government* (Australia). Extracts from Concepts and Techniques. Programme Budgeting; Strategy for Continuing Development.,Department of Management and Budget Program Development and Review Division. 125 pages.

World Bank, 1993 *Water Resources Management* , a World Bank Policy Paper. Washington D.C.

World Bank. 1997. *Rural Development:From Vision to Action.* A Sector Strategy. Washington D. C. 187 pages.

World Bank/UNDP. 1990. *Irrigation and Drainage Research*: A Proposal for an Internationally-Supported Program to Enhance Research on Irrigation and Drainage Technology in Developing Countries. 21 pages.

T - #0708 - 101024 - C0 - 276/216/9 - PB - 9789054104834 - Gloss Lamination